Digital Futures in Human-Computer Interaction

The application of futures thinking in Human-Computer Interaction (HCI) has become increasingly important in recent years. Integrating speculative thinking with future design approaches has allowed HCI researchers to explore the potential impacts of technology on digital society. However, the implementation and application of futures thinking in HCI research is an emerging area. *Digital Futures in Human-Computer Interaction: Design Thinking for Digital Transformation* fills this gap by systematically analyzing HCI's innovation trends in the digital era.

This book explores the dialogue between digital transformation and futures thinking for alternative visions of HCI research. The book highlights significant trends and advancements in futures thinking related to HCI. Case studies illustrate the role of futures thinking, offering readers a broad overview of the subject while detailing the competencies and practices that can lead to successful futures design.

This engaging and informative reference will appeal to students, academics, and researchers interested in various design aspects related to HCI. These aspects include service design, sustainable design, product design, space design, visual communication, design education, futures studies, and social innovation.

Digital Futures in Human-Computer Interaction

Design Thinking for Digital Transformation

Edited by

Zhiyong Fu, Anna Barbara, and Peter Scupelli

Designed cover image: image credited to PST Vector [ShutterShock ID: 2315568623]

First edition published 2025
by CRC Press
2385 NW Executive Center Drive, Suite 320, Boca Raton FL 33431

and by CRC Press
4 Park Square, Milton Park, Abingdon, Oxon, OX14 4RN

CRC Press is an imprint of Taylor & Francis Group, LLC

Library of Congress Cataloging-in-Publication Data
Names: Fu, Zhiyong, editor. | Barbara, Anna, editor. | Scupelli, Peter, editor.
Title: Digital futures in human-computer interaction : design thinking for digital transformation / eited by Zhiyong Fu, Anna Barbara and Peter Scupelli.
Description: First edition. | Boca Raton, FL : CRC Press, 2025. | Includes bibliographical references and index.
Identifiers: LCCN 2024031554 (print) | LCCN 2024031555 (ebook) | ISBN 9781032607603 (hardback) | ISBN 9781032693583 (paperback) | ISBN 9781032693606 (ebook)
Subjects: LCSH: Human-computer interaction.
Classification: LCC QA76.9.H85 D494 2025 (print) | LCC QA76.9.H85 (ebook) | DDC 004.01/9--dc23/eng/20241129
LC record available at https://lccn.loc.gov/2024031554
LC ebook record available at https://lccn.loc.gov/2024031555

ISBN: 978-1-032-60760-3 (hbk)
ISBN: 978-1-032-69358-3 (pbk)
ISBN: 978-1-032-69360-6 (ebk)

DOI: 10.1201/9781032693606

For Product Safety Concerns and Information please contact our EU representative:
GPSR@taylorandfrancis.com
Taylor & Francis Verlag GmbH,
Kaufingerstraße 24,
80331 München, Germany.

Contents

Foreword

In the 1980s, a group of futurists (James Dator, Lestor Cingcade, Wayne Yasutomi, and myself) explored the futures of robotics/artificial intelligence (AI). As we were working with the Hawaii Judiciary (Cingcade himself was the Administrative Director), we imagined a future where lawyers and judges would first increasingly use AI for minor judicial decision-making (administrative tasks such as parking fines) and eventually adopt it for more complex judicial decision-making. We then asked, what would be the conditions where AI/robotics would become so ubiquitous that robots would have legal rights. Certainly, most judicial leaders could not see the relevance of our speculations. However, they did understand that institutions needed futures literacy to stay relevant to the changing needs of citizens. Thinking of the unthinkable – robotic rights – distances us from the present, allowing us to see the present as not only remarkable, but designable. It spurs creativity and ensures that the outlier is embraced.

Futures literacy is not only about seeing the new possibilities in emerging innovation but also using the future to redesign the present. To think, feel, and act differently. To do this, futurists first challenge the used future – current institutional practices that no longer work but continue because we have always done it that way. Once this used future is challenged, we then anticipate and create alternative futures.

This novel book will help in this task. It explores how the future of human-computer interaction is likely to change and what policies – frameworks, ethical guidelines, narratives – are needed to guide this change. Evolution is moving from random to co-designed: humans with nature with technologies. In this process, merely living the programmed futures of others will not suffice. We need to innovate and co-design with actors and networks, human and non-human to create a transformative future. This book can help us in this process.

Sohail Inayatullah
UNESCO Chair in Futures Studies

Foreword

Preface

In an era marked by rapid technological advancements, digital transformation has become a cornerstone of innovation across industries. Human-Computer Interaction (HCI), as a field where technology and human interaction intersect, is at the forefront of this transformation, facing unprecedented opportunities and challenges. Within this fast-evolving landscape, futures thinking has emerged as an essential tool to drive innovation in HCI, providing a new way of thinking that empowers us to transcend the present, envision, and shape future technological ecosystems and societal structures. Futurescaping, a central concept in futures thinking, is not merely a tool for prediction but a design philosophy. By leveraging imagination, fiction, and extrapolation, the HCI field is poised to explore new possibilities and break through the limitations of current technological and social paradigms.

This book delves into the dynamic relationship between digital transformation and futures thinking, providing a comprehensive analysis of future trends, futures literacy, and practical applications. It aims to illuminate how these strategies can help shape preferable futures in the HCI domain, involving not only technological innovation but also deep reflection on societal impacts. As HCI evolves, there is a growing responsibility to ensure that the technologies we create contribute to a sustainable and human-centered future.

The book is structured to offer both theoretical insights and practical tools for those interested in the future of HCI. Section I establishes the foundation by examining future trends in HCI. It opens with an analysis of the effects of digital transformation on industries such as finance, healthcare, and education, and then explores emerging technologies in areas like care services and museums. Section II delves into futures literacy, examining methodologies for futures thinking, exploring pedagogical approaches for entertainment technology, and emphasizing long-term thinking in digital product-service systems, especially for small and medium-sized enterprises (SMEs). Section III shifts focus to practical applications and competencies, covering current methods in design futures, AI empathy in the metaverse, speculative design artifacts, interactive narratives, and affective evaluation strategies in human-AI interactions. Section IV concludes with a discussion of the broader societal implications of these emerging trends. It explores sustainability-oriented design futures methods, ecological thinking in HCI, the societal impacts of third-wave HCI research, and the influence of digital technology on personal well-being and quality of life.

Throughout the book, a careful balance is struck between theoretical exploration and practical application. By focusing on both technological innovation and human-centered design, it addresses the evolving challenges in HCI while anticipating future possibilities. Case studies and real-world examples enrich the discussion, providing a comprehensive understanding of how futures thinking can guide HCI research and practice.

As digital transformation continues to accelerate, HCI will face increasingly complex challenges and opportunities. Futures thinking offers the critical framework needed to guide technological development toward more sustainable and ethically responsible outcomes. By encouraging reflection on current practices and thoughtful projection into the future, we can shape a more inclusive and balanced technological future. In particular, areas such as artificial intelligence and the metaverse present significant opportunities, but they also require careful consideration of their long-term social and ecological impacts.

This book is a valuable resource for anyone seeking to explore the intersection of technology, design, and the futures that lie ahead. It invites readers from diverse design disciplines—including service design, product design, and social innovation—to harness the power of futures thinking in shaping a better tomorrow, where technology and humanity evolve together in harmony.

Zhiyong Fu, Anna Barbara, and Peter Scupelli
Co-founders of Global Design Futures Network (GDFN)

Authors

Zhiyong Fu is a tenured associate professor and doctoral supervisor at the Department of Information Art and Design, Academy of Arts and Design, Tsinghua University. He serves as the Director of the Art and Technology Innovation Base at Tsinghua University, Deputy Director of the China Innovation and Entrepreneurship Education Research Center, Secretary-General of the Information and Interaction Design Committee of the China Industrial Design Association (IIDC) President of the International Chinese Association of Human-computer Interaction (ICACHI). He is Co-founder of the Global Design Futures Network (GDFN), the member of the Second Global Important Agricultural Heritage Systems Expert Committee of the Ministry of Agriculture and Rural Affairs and a Director of the China-Europe Alliance for Arts and Humanities Education.

His teaching responsibilities include undergraduate and graduate courses in interaction design, service design, design thinking, and innovation and entrepreneurship. His research interests encompass intelligent product and service design, smart cities and design futures. He is the leader for multiple national and international research projects and published papers in key international conferences/journal in HCI and design. Fu was a recipient of the Ministry of Education's New Century Excellent Talents Support Program in 2006 and was a visiting scholar at the University of Tsukuba, Japan, in 1998, and Carnegie Mellon University, USA, in 2008. He played a key role in planning and organizing the China-U.S. Young Maker Competition under the Ministry of Education from 2014 to 2023 and he currently co-leads GDFN, a future-oriented global creative community that is preparing future-thinking innovators to tackle sustainability challenges with the design futures methodology.

Anna Barbara is Associate Professor in Architecture and Interior Design at the Design Department, Politecnico di Milano; President of POLI.design; Member of the Board of Directors of the World Design Organisation; Co-founder of the Global Design Futures Network; Scientific coordinator (with Venere Ferraro) of the D\Tank, Design Department, Politecnico di Milano; Italian Design Ambassador (2020, 2021, Kuwait; 2022, Sudan). Graduated in Architecture at Politecnico di Milano, she is teaching as visiting at Tsinghua University, Academy of Art and Design, Beijing (2019, 2024, China); at Kookmin University, School of Architecture, Interior Design and at Master Brain 21 (1999, South Korea). She was Canon Foundation Fellow at Hosei University (2000, Japan), Special Mention of Borromini Prize (2001), Selected by Archmarathon (2018), Selected ADI-Index 2019, 2023, Special Mention Fedrigoni Top Award – Large Format Communication, 2023; Eccellenze della Lombardia 2019, 2023; Exhibited t Biennale di Venezia (2010, 2022), Triennale di Milano (2003, 2016). Author of Storie di Architettura attraverso i sensi (2000, 2020), Invisible Architectures. Experiencing places through the senses of smell with Anthony Perliss (2006, 2023); Sensi, tempo e architettura (2012), Sensefulness, new paradigms for Spatial Design (2019), Extended Store. How digitalization affects the retail space design with Yuemei Ma (2021). She designed international sensorial projects for companies in Japan, China, USA, Europe, UAE, and UK as founder of Senselab.

Peter Scupelli is Associate Professor of Design, and Director of the Learning Environments Lab. He was Visiting Professor at Politecnico di Milano (2022-2024) and the Nierenberg Professor at Carnegie Mellon University (2019-2022). Peter's current research focus is on learning environments and sustainable design futures. Peter is co-founder of the Global Design Futures Network (GDFN). Peter was an executive co-chair for the International Conference for Design Futures (ICDF) from 2020 to 2023. He holds a Ph.D. in Human-Computer Interaction, M.S. in HCI, M.Des. in Interaction Design, and an undergraduate Architecture degree. Peter has published over fifty-seven peer-reviewed papers at international conferences and journals. His research has been funded by the National Science Foundation and Institutes of Educational Science.

Peter teaches undergraduate, masters, and doctoral students. His teaching and research focus on three fundamental topics necessary to bring aspects of Transition Design to design practice: (a) aligning short-term design action and long-term vision goals, (b) embedding values into design processes, and (c) creating sustainable futures-oriented design methods. In *Dexign Futures*, students learn to combine Design Thinking with Futures Thinking to align short- and long-term time horizons. In *Design for Zero-Carbon Lifestyles*, he teaches how to live a zero-carbon lifestyle and create zero-carbon organizations. In *Design Your Futures*, students learn to design their life.

His work with Gruppo A12 was exhibited in the Architecture Biennale of Venice; PS1-MOMA, New York; the São Paulo Contemporary Art Biennial; the ZKM Museum of Karlsruhe, Germany, and many other places.

Contributors

Anna Barbara
Department of Design, Politecnico di Milano
Milano, Italy

Chiju Chao
Academy of Art and Design, Tsinghua University
Beijing, China

Harshit Desai
MIT Institute of Design, MIT ADT University
Pune, India

Venere Ferraro
Department of Design, Politecnico di Milano
Milano, Italy

Zhiyong Fu
Academy of Arts & Design, Tsinghua University
Beijing, China

Nandhini Giri
Department of Computer Graphics Technology, Purdue University
West Lafayette, Indiana

Shams Hamid
IMAGINE - Institute of Futures Studies
Karachi, Sindh, Pakistan

Yasuyuki Hayama
Department of Strategic Design, Faculty of Design, Kyushu University
Fukuoka, Japan

Jiawei Li
Academy of Arts & Design, Tsinghua University
Beijing, China

Yin Li
Academy of Arts & Design, Tsinghua University
Beijing, China

Zhicong Lu
City University of Hong Kong
Hong Kong SAR, China

Yanru Lyu
Department of Digital Media Arts, Beijing Technology and Business University
Beijing, China

Xiaojuan Ma
Department of Computer Science and Engineering, the Hong Kong University of Science and Technology
Kowloon, Hong Kong

Peter Scupelli
School of Design, Carnegie Mellon University
Pittsburgh, Pennsylvania

Erik Stolterman
School of Informatics, Computing, and Engineering, Indiana University
Bloomington, Indiana

Jiayue Wang
Academy of Arts & Design, Tsinghua University
Beijing, China

Qing Xia
School of Humanities (Dalian), Luxun Academy of Fine Arts
Dalian, Liaoning, China

Bowen Zhang
Department of Digital Media Arts, Beijing Technology and Business University
Beijing, China

Lin Zhu
Academy of Arts & Design, Tsinghua University
Beijing, China

Section I

Future Trends in HCI

1 Digital Transformation and Future HCI

Zhiyong Fu and Jiawei Li

INTRODUCTION

Digitalization and Digital Transformation: An Analysis Framed by Actor-Network Theory

Background of Digital Transformation

In the 21st century's wave of digitization, digital transformation has become the engine driving profound changes in the global economy and society. This transformation not only involves the comprehensive reshaping of enterprise operating models and work modes but also touches upon various aspects of people's daily lives. The pervasive nature of digital transformation is evident not only in reshaping business ecosystems and market landscapes but also in exerting profound influences on cultural forms and social interactions. For example, the rise of digital media and social platforms has massively altered the ways people acquire information and engage in communication, while the widespread application of big data and artificial intelligence (AI) technologies is gradually redefining operational standards and productivity boundaries across various industries.

Tratkowska's (2019)'s paper analyzes the critical role of the VUCA (Volatility, Uncertainty, Complexity, and Ambiguity) environment in digital transformation and explores strategies to address these challenges. In the VUCA era full of volatility, uncertainty, complexity, and ambiguity, organizational digital transformation has become a necessary step to gain competitive advantage and ensure survival (Mukhlisah, 2023). Digital transformation serves not only as a potent driving force for the development of organizations worldwide but also plays a pivotal role in strategic formulation, operational efficiency enhancement, business model innovation, and employee capability improvement (Çebi & Gözlü, 2023). Thus, it further suggests that digital transformation significantly affects the ability to alleviate financing constraints, enhance the identification, and capture of innovation opportunities, thereby helping reduce the risk of enterprise innovation investment and improving the level of investment volatility management. Under the impetus of the digital wave, organizations can not only better adapt to changes in the external environment but also achieve more precise market positioning, more efficient resource allocation, and more flexible business model innovations through the exploration and utilization of data resources. For organizations, thoroughly understanding and seizing the opportunities and challenges of digital transformation and actively responding to the uncertainty brought by the VUCA environment will be crucial to future development. By formulating scientific digital transformation strategies, optimizing organizational

DOI: 10.1201/9781032693606-2

structures and processes, and enhancing employee digital literacy, organizations can establish themselves firmly in the digital age and achieve sustainable development.

The theoretical background of digital transformation is extensive and in-depth, covering multiple key areas. Bonanomi points out in his research that multidisciplinary design firms, when promoting digital transformation, need to consider multiple dimensions such as collaboration, iteration, platform, and network paradigms (Bonanomi & Bonanomi, 2019). These paradigms not only integrate considerations at the process and organizational levels but are also influenced by the unique characteristics of the company and its environment. In addition, Xu, Hou, and Wang delve into the frontier research of digital transformation and construct a comprehensive and systematic theoretical framework (Xu et al., 2022), aiming to provide strong support for management practices and guide future theoretical explorations and empirical studies in this field. In the domain of digital workplaces, Weritz emphasizes the importance of key elements such as entrepreneurial mindset, digital responsibility thinking, digital literacy, and change skills (Weritz, 2022). He believes that these capabilities are essential for individuals' development and growth in the digital age, including skills in communication, community management, data analysis, and web development. Markus and Rowe point out in their research that in order to fully utilize diverse theoretical representations of digital transformation, the information systems field needs to clearly distinguish and accurately label various phenomena (Markus & Rowe, 2023). Finally, the research by Hanelt, Bohnsack, Marz, and Antunes further reveals the impact of digital transformation on organizational design (Hanelt et al., 2021) They point out that digital transformation endows organizational design with plasticity, enabling it to continuously adapt to changing environments. This process is driven by digital business ecosystems and is to some extent covered by traditional organizational change frameworks.

Digital transformation is not only a technological innovation but also a profound change in business and organizational models. Kamila Tratkowska's definition in the journal "Management Sciences." reveals its essence: digital transformation is far more than mere technological applications; it touches the very foundation of business models, including value creation, customer interaction, and the reshaping of internal processes (Tratkowska, 2019). At the core of this transformation is the use of digital technology to drive these deep-seated changes, thereby enhancing efficiency, improving customer experience, and exploring new market opportunities. This also provides diverse paths and strategic choices for numerous organizations in pursuit of digital goals, highlighting the complexity and multidimensionality of digital transformation. With the deepening of digital transformation, its impact has permeated various industries, from automotive, energy, and manufacturing to healthcare, banking, tourism, and education, all undergoing profound changes (Haktanır et al., 2022). In the public sector, digital transformation can bring about experimental, inclusive, and anticipatory governments, promoting innovation and creating public value (Kaivo-oja et al., 2022). It is worth noting that some "digital-native" companies such as Alibaba, Uber, and Airbnb, propelled by the winds of digital transformation, have reshaped traditional industry landscapes through innovative business models. By leveraging digital transformation, they break through the boundaries and limitations of traditional industries, providing consumers with more convenient, efficient, and personalized service experiences. The successful practices of these enterprises not only offer valuable experiences and insights for other businesses but also further demonstrate the important role of digital transformation in driving industry innovation and development.

Meanwhile, the rapid development of digital technologies such as robotics, AI, and virtual reality has deeply integrated into our daily lives, altering our perceptions and modes of interaction (Cinque & Vincent, 2022). In this context, digital twin technology has emerged, bringing new possibilities to organizations, production, and life matrices. The application of digital twin technology in the organizational, production, and life matrices can span multiple dimensions of this matrix. Digital twins are a technology for creating virtual replicas of physical entities, allowing real-time simulation and analysis of the state and behavior of entities. At the organizational level, digital twins can be used to optimize operations, enhance decision-making, and improve efficiency.

Actor-Network Theory

Actor-network theory (ANT) is a theoretical framework for studying the interaction between society and technology, first proposed by Bruno Latour and others in the 1980s (Latour, 2007). The idea originated from the early 1980s at the Centre de Sociologie de l'Innovation (CSI) of the École des Mines in Paris, developed by Michel Callon, Bruno Latour, and John Law, reflecting a post-structuralist concern with non-foundationalism and the multiplicity of material-semiotic relations. At its core, ANT posits that society is composed not only of human actors but also includes non-human actors (such as technologies, objects, concepts, etc.) playing crucial roles within networks. This theory emphasizes the relationships between actors and the process of network construction, suggesting that the formation and development of socio-technical systems occur through ongoing interactions and negotiations among actors. It provides a method for understanding and analyzing the complex interactions within systems, particularly suitable for examining phenomena like digital transformation that are multidimensional and cross-disciplinary. ANT can be viewed as an analytical framework for studying the interactions among science, technology, and society, aiding in understanding processes of technological development and knowledge creation. This approach challenges purely social or technological explanations, emphasizing the relationships between social and technical elements, whether human or non-human (Wikipedia contributors, n.d.) Characterized by its relational, processual, and constructivist nature, ANT can be employed to investigate scientific practices and technological developments. The theory has been widely applied across various disciplines and fields, often serving as implicit background knowledge rather than an explicit theoretical or methodological paradigm.

Within the context of digital transformation explored in this chapter, ANT offers a unique perspective by conceptualizing digital transformation as a network of actors including technologies, human users, developers, regulatory bodies, as well as social norms and practices. It underscores the agency of non-human actors, such as software and algorithms, which is crucial for understanding advancements in human-computer interaction (HCI). Moreover, the concepts of translation and negotiation within ANT are essential for understanding the interactions among various stakeholders involved in digital transformation. Additionally, ANT's perspective of heterogeneous engineering aids in comprehending the design and development of digital platforms that integrate diverse user needs and technological possibilities. Through this theoretical framework, this chapter explores how "key actors" in digital transformation, such as technologies, humans, organizations, etc., interact and how these interactions form networks that either facilitate or impede transformation. By analyzing trends in HCI development and technological innovation cases in key industries such as finance, retail, healthcare, education, and manufacturing, it demonstrates how the integration of design thinking, future thinking, and AI technology is driving rapid advancements in the HCI field.

In summary, ANT provides robust theoretical support for understanding and guiding the development of digital transformation and future HCI. By revealing the complex interactions among technology, users, and society, ANT helps us better understand how to design and implement more effective and user-centric HCI systems.

An ANT Perspective on Digital Transformation and HCI

To gain a deeper understanding of how digital transformation influences and shapes HCI development, we can employ the perspective of ANT. ANT emphasizes the interactions between human and non-human actors, as well as the importance of translation and negotiation processes in shaping network evolution. Figure 1.1 illustrates a network diagram based on ANT theory, demonstrating key actors and their interactions in digital transformation.

In this network, human actors and non-human actors (such as technologies, business processes, organizational structures, etc.) are intertwined, collectively driving the process of digital transformation. Human actors influence non-human actors through technological innovation, changes in business models, and restructuring of organizational patterns, while non-human actors shape the decisions and behaviors of human actors through empowerment and constraints. This bi-directional interaction reflects the mutual shaping relationships among actors in ANT theory. Various elements of digital transformation, such as technological innovations like AI, big data analytics, cloud computing, Internet of Things (IoT), as well as business and organizational changes like value creation, customer interactions, internal process optimization, and employee skill enhancement, all interact through translation and negotiation processes, ultimately shaping the direction of HCI development. Simultaneously, factors within the field of HCI, such as design thinking, future thinking, and social impacts, also interact with various elements of digital transformation through translation and negotiation processes, influencing the outputs and outcomes of HCI, such as user insights, rapid iterative development, seamless interaction experiences, personalized services, and social responsibility considerations. By presenting a complex interaction network between digital transformation and HCI based on ANT theory, this chapter highlights the critical roles of both human and non-human actors in translation and negotiation processes, as well as how these processes collectively drive the development and evolution of HCI. This offers a new perspective for comprehensively understanding the impact of digital transformation

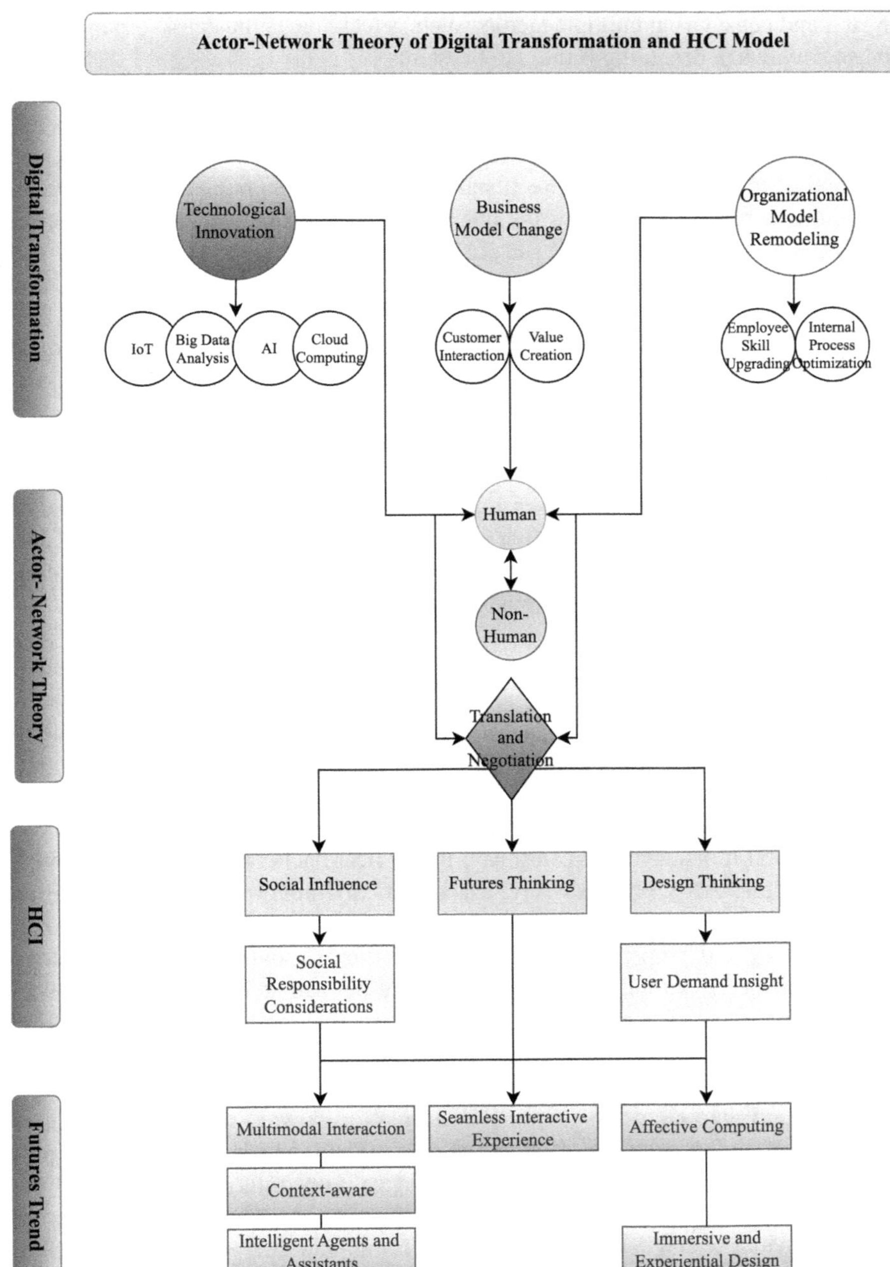

FIGURE 1.1 An ANT Network of Digital Transformation and HCI

on HCI and considering multiple factors, such as technology, business, organization, and society, when designing future HCI systems.

To further elucidate and analyze the relationship between digital transformation and HCI, the authors introduces the concepts of "human actors" and "non-human actors" from ANT theory and represents their interactions in digital transformation using subgraphs. Emphasizing the bi-directional influence between human and non-human actors, this approach underscores the interactions and mutual shaping among actors in ANT theory. In the section on HCI, the author introduces the "translation and negotiation" processes, linking factors such as design thinking, future thinking, and social impacts, demonstrating how they collectively influence the development of HCI through translation and negotiation. The outputs of HCI, including user insights, rapid iterative development, seamless interaction experiences, personalized services, and social responsibility considerations, are all formed through translation and negotiation processes between human and non-human actors, further highlighting the core concepts of ANT theory, such as the interaction between human and non-human actors, translation and negotiation processes, and how these processes collectively shape the development of digital transformation and HCI. Additionally, it illustrates the close connection between digital transformation and HCI and how various factors influence HCI outputs and outcomes through complex interactions.

Factors Driving Digital Transformation and ANT Analysis

Digital transformation represents a critical challenge and opportunity for contemporary societies and organizations. It involves leveraging digital technologies to reimagine and redesign business models, products, and services in response to rapidly changing market demands and societal expectations (Vial, 2021). Applying ANT to analyze digital transformation can unveil the intricate networks of actors involved. In this process, ANT provides a robust analytical framework for understanding how technology, humans, and organizations interact within the socio-technical networks shaping and propelling digital transformation (Latour, 2005a). Technology is no longer merely a tool or medium for change but can actively participate as an actor in the construction of socio-technical systems. For instance, cloud computing platforms, as non-human actors, interact with enterprises (human actors) and market strategies (conceptual actors), jointly driving innovation and evolution in business models. Enterprises, by adopting cloud platforms, gain more flexible access to, storage of, and processing of large volumes of data, which may have been previously unattainable due to cost and technological constraints. This adoption not only alters the IT infrastructure of enterprises but also facilitates the emergence of new work patterns and service models. Within this new socio-technical structure, the interaction between cloud platforms and enterprises fosters process optimization, improved collaboration efficiency, and the creation of new services. Similarly, cloud platforms are closely linked with conceptual actors like market strategies. With intensified market competition, enterprises need to respond more flexibly and agilely to market changes, and the elasticity and scalability of cloud platforms meet this demand. Enterprises utilize cloud platforms to deploy applications and services rapidly, achieving personalization and customization, thereby better attracting and retaining customers. Additionally, cloud platforms interact with regulatory bodies in the form of government regulations and standards. With the strengthening of data security and privacy regulations, cloud service providers must ensure compliance with legal requirements. This may influence the design and management of cloud services, thereby impacting how enterprises utilize these services to support their operations. Ultimately, consumer behavior also acts as a significant actor influencing the application and development of cloud platforms. As user expectations for internet services continue to rise, they increasingly favor services that offer seamless, instantaneous, and personalized experiences. Cloud platforms enable enterprises to meet these expectations, launching more attractive products and services, thereby shaping consumer usage habits and preferences. In this complex network comprised of diverse actors, cloud platforms symbolize not only technological innovation but also crucial participants driving socio-economic transformation and innovation in the digital age. Through interactions with other actors, cloud platforms continuously reshape the business landscape of the digital era, demonstrating their significant role in contemporary socio-technical change.

ANT challenges traditional technological determinism and social constructivist perspectives, proposing that technology and society are co-constituted through interactions within networks. This perspective is particularly useful for analyzing digital transformation as it necessitates researchers to consider a wide range of actors, including users, developers, policymakers, and the technology itself, and how they collaborate and compete within evolving networks. Furthermore, digital transformation is also regarded as a key driver of social and economic development. Kallinikos, Aaltonen, and Marton explore how digital technologies reshape modes of information processing, communication, and organizational collaboration, thereby influencing social structures and individual behaviors (Kallinikos et al., 2013). These changes not only reflect technological advancements but also embody a reimagination of how people live and work. Finally, as digital transformation deepens, considerations for ethics and privacy become increasingly important. The data collection and processing activities spurred by digital transformation exacerbate concerns regarding individual privacy protection and data security (Zuboff, 2019). Within the analytical framework of ANT, these concerns are viewed as actors within the network, whose voices and rights need to be fully considered and protected throughout the digitization process. By applying the core concepts of ANT to analyze digital transformation, we can gain a deeper understanding of how various actors interact and how these interactions shape and reshape the trajectory and outcomes of digital transformation. The author believes that ANT provides a robust framework for exploring and analyzing the complexity and dynamism of digital transformation, emphasizing the roles of both human and non-human actors in shaping and developing digital networks (Table 1.1).

KEY INDUSTRY EXAMPLES OF DIGITAL TRANSFORMATION

Digital transformation has brought revolutionary changes to traditional industries such as finance, retail, healthcare, and education by introducing digital technologies like mobile applications, online platforms, and AI, significantly enhancing

TABLE 1.1
ANT Core Concept Explanation and Application in Digital Transformation

Core Concepts of ANT	Conceptual Explanation	Application in Digital Transformation
Actor	Actors can be individuals, groups, organizations, or any non-human entities capable of influencing the network. In ANT, actors have equal status, whether human or non-human.	In digital transformation, actors include employees, management, IT systems, smart devices, and so on. For example, smart devices (such as IoT devices) act as actors by collecting and analyzing data, influencing business decisions and operations.
Network	The network is a relational structure formed by interactions among actors. These relationships define how actors influence each other and the behavior and development of the entire network.	The networks involved in digital transformation encompass various interconnected elements such as technological infrastructure, organizational structure, and business processes. For instance, cloud computing platforms connect different business applications and data storage, forming networks that support digital businesses.
Translation	Translation refers to the process of interaction among actors, through which actors translate their intentions, needs, or goals into actions and relationships within the network.	Translation occurs during the process of transforming organizational goals into specific technical solutions. For example, the digital strategy of a company (the intent of the actors) is translated into concrete technology deployments and business process transformations.
Mediator	Mediators in ANT refer to entities that, during the process of transmission, can change, transform, or reconstruct the content they transmit. Mediators cause changes in relationships and actions within the network.	In digital transformation, software tools and platforms serve as mediators; they not only transmit information but also can change work methods and business models. For example, data analytics software can transform raw data into insights, thereby altering the decision-making process.

operational efficiency and user experience, and driving service modernization. The author discusses digital banking and digital e-commerce as key examples.

In the financial sector, Singapore's DBS Bank has successfully transformed into a leading digital bank through its "digital banking" strategy, introducing mobile banking apps, online trading platforms, and AI customer service, greatly optimizing traditional banking service models (Garg et al., 2021). In its digital transformation process, DBS adopted a "comprehensive and progressive" strategy. On the one hand, it vigorously embraced emerging digital technologies, continuously innovating digital products and services through a combination of internal development and external collaboration. On the other hand, it emphasized integration with traditional businesses, striving for seamless integration of online and offline, and old and new systems. Specifically, DBS invested heavily in digital infrastructure construction, building a cloud-native modern IT architecture, significantly improving system flexibility and response speed. It also redesigned the entire digital operation and service process, improving front-end service efficiency and reducing back-end operating costs. In terms of customer experience, DBS is committed to providing users with simple, intelligent, and seamless experiences. By analyzing customer behavior preferences through big data, it launches personalized products and services. Leveraging technologies such as AI and machine learning (ML), it provides value-added services such as intelligent risk control and intelligent investment advice. Meanwhile, DBS Bank's integrated approach, combining its mobile app, online banking, and physical branches, offers customers a consistent cross-channel experience. Additionally, DBS actively embraces the concept of open banking, collaborating with fintech companies, e-commerce platforms, and social media to innovate financial scenarios continuously, expanding user touchpoints. For instance, it introduced the PayLah! e-wallet, integrating into daily life payment scenarios, and partnered with Taobao to launch the Tao Silver service, among others. These digital initiatives helped DBS Bank achieve low-cost, high-efficiency operations while providing customers with unprecedented premium experiences, facilitating its successful transformation into a leading digital bank.

In the retail industry, e-commerce giants Alibaba and Amazon have reshaped the shopping experience and supply chain management through digital platforms. Alibaba's "new retail" strategy, in particular, integrates online and offline scenes through an O2O model, delivering a seamless shopping experience to consumers and fundamentally changing traditional retail formats. Specifically, Alibaba, through its platforms such as Tmall and Taobao, has built an efficient digital trading scene for consumers and merchants. Consumers can purchase and pay online, enjoying convenient shopping experiences, while merchants utilize the platform for precise marketing and channel expansion. Alibaba also built an intelligent logistics system to improve delivery efficiency (Kim, 2018). Moreover, Alibaba's "new retail" strategy further integrates online and offline businesses, upgrading physical store operations through digital means to provide consumers with a seamless experience. On the other hand, Amazon's success stems from its advanced digital supply chain system and customer-centric philosophy. Through big data analysis, Amazon can accurately predict customer demands and achieve efficient inventory management. Its core lies in providing rich, low-cost, and efficient products and services. Additionally, Amazon has vigorously developed its self-operated logistics and AI technologies to continuously enhance operational efficiency. The digital practices of these two e-commerce giants greatly enrich consumers' shopping scenarios, bring new business growth opportunities to merchants, and drive the entire retail format toward digitization and intelligence.

During the pandemic, telemedicine has become an important means to ensure public health through digital means such as mobile applications, video consultations, and AI assistance, providing convenient medical services to patients at their doorsteps (Smith et al., 2020). In the education sector, online education platforms such as Coursera, Khan Academy, and edX have emerged, breaking geographical and time constraints, making high-quality educational resources accessible, and meeting the needs of personalized and flexible learning for students. It is evident that digital transformation empowers traditional industries comprehensively, driving efficiency improvements and service optimization, bringing unprecedented convenience and intelligent experiences to people's production and life.

THE DEVELOPMENT TRENDS OF HUMAN-COMPUTER INTERACTION: STARTING FROM THE ANT PERSPECTIVE

HCI is a discipline that studies the interaction between humans and computer systems, aiming to design more user-friendly, efficient, and humane technological products and services. With the deepening of digital transformation, the HCI field faces unprecedented challenges and opportunities. Combining the characteristics of ANT mentioned earlier provides a useful analytical perspective for understanding and deciphering the intricate interactions between technology and human actors in HCI.

First, ANT emphasizes that HCI involves not only direct interactions between humans and machines but also the roles of various non-human actors such as the environment, society, and culture. The work of Norman emphasizes the importance of design in optimizing HCI, proposing that design principles should fully consider human needs and limitations (Norman, 2013). Through the theoretical framework of ANT, we can further expand this viewpoint, viewing the design process as an interaction and negotiation among human and non-human actors such as machines, environments, societies, and cultures. Suchman, through case studies of HCI, proposed the concept of "situated action," emphasizing the situational and interactive nature of the interaction process (Suchman, 1987). By applying ANT theory, this insight can be extended to a broader social-technical network, thus understanding how HCI is a dynamically constructed process involving multiple actors. On the other hand, with the continuous evolution of technology, the HCI field is exploring more complex interaction methods, such as communication with computer systems through touch, sound, brain signals, etc. Kaptelinin and Nardi discussed the application of activity theory in HCI design, emphasizing the interactive relationship between human activity practices and technology (Kaptelinin & Nardi, 2012). Combining activity theory with ANT provides an integrated theoretical framework for analyzing and designing complex HCI systems. Looking back to the present, the rise of AI and ML technologies has brought new changes to HCI. Oulasvirta and Hornbæk studied the role of AI in HCI design, particularly how AI is used to predict and adapt to user needs and behaviors (Oulasvirta & Hornbæk, 2016). From the perspective of ANT, the progress of such technology reflects not only the evolution of technology itself but also the redefinition of the roles and relationships between human and machine actors in the constantly changing social-technical network. Overall, with the continuous advancement of digital technology, ANT will undoubtedly open up new theoretical perspectives and practical paths for HCI research. ANT provides a social-technical integrative analysis perspective, helping to comprehensively understand the interactions among various actors in the HCI process and the relationship between technology and human activity practices, thus supporting the design of more humane and situational interaction experiences.

Four Waves of HCI

HCI is a multidisciplinary field focused on understanding and designing interactions between humans and computers. It integrates theories and methods from various disciplines such as anthropology, sociology, cognitive science, and psychology. The scope of HCI technology covers various aspects, from ethnography and participatory design to usability testing and controlled experiments. Over the years, HCI has evolved to involve various aspects of people's lives, including spirituality, global crises, disabilities, as well as emerging technologies such as wearable devices, robots, and virtual reality. The future prospects of HCI research include gamified interactions, self-tracking and behavior change technologies, conversational agents, among others (Seaborn & Fels, 2015; Kersten-van Dijk et al., 2017; Weger & Yeazitzis, 2023). The development of HCI can be divided into four stages or waves: the first wave of HCI focused on cognitive science and human factors, employing strict guidelines, formal methods, and systematic testing; the second wave shifted toward interactions within work environments and practice communities, introducing concepts such as context and adopting active methods like participatory design workshops and contextual inquiry (Bødker, 2015); the third wave emphasized the contextual nature of HCI, recognizing that interactions are embedded in complex and constantly changing environments, and emphasized the staged nature of language communication (Borin & Edlund, 2018). Some scholars advocate for embracing the fourth wave of HCI, prioritizing politics, values, and ethics, driving true institutional change, and incorporating political activism into mainstream HCI (Rydenfält & Persson, 2020). Additionally, feminist theories of science, technology, and society are used to examine normative cultures in computer science, revealing gender differences and power relations in the field (Ashby et al., 2019).

What Are the Challenges and Opportunities for HCI in the Next Wave?

Currently, HCI is in a new era full of opportunities and challenges. With the rapid development of emerging technologies such as AI, the IoT, and virtual reality, HCI needs continuous innovation and adaptation to meet the growing needs and expectations of people. In this process, HCI faces many challenges but also contains enormous opportunities.

First, the emergence of human-computer hybrid systems poses new challenges to HCI. In these systems, humans and machines need to collaborate seamlessly to complete tasks. However, how to ensure effective communication and coordination between humans and machines, and how to manage and optimize the accuracy of language interactions are urgent issues to be addressed (Dove et al., 2022). Additionally, creating and conveying personalized interaction experiences in dialogue systems is also a topic worthy of in-depth exploration (Pinhanez, 2020). At the same time, mastering the appropriate timing of interactions, knowing when to initiate dialogue actively and when to listen silently, is crucial for enhancing user experience (Frauenberger, 2019). Despite the many challenges, HCI also faces numerous opportunities in the next wave of development. With the popularity of the IoT, supporting interactive skill sharing and autonomous learning becomes possible. Users can share their skills with other users or smart devices in the IoT environment, or learn new skills autonomously by observing and imitating (Jones et al., 2023). This new interaction mode is expected to greatly enhance user creativity and productivity. Another opportunity is the development of Entanglement HCI. This emerging field focuses on the performance relationship between humans and technology, reconstructing the process of knowledge generation, tracking accountability and ethical issues, exploring design and material practices, etc. By integrating humanities and social sciences with engineering technology, Entanglement HCI is expected to reveal the deep-seated rules of HCI, and promote HCI toward a more humane, intelligent, and trustworthy direction. In addition, HCI education also faces opportunities for change and innovation. By studying the interaction modes between students and users, exploring reflective teaching methods, we can cultivate more HCI talents with interdisciplinary perspectives and innovative capabilities (Roldan et al., 2020). These talents not only master technology but also have profound humanistic literacy and social responsibility, able to think and design from the perspective of users, promoting HCI to better serve the development of human society.

The author believes that HCI is both facing challenges and full of opportunities in the next wave of development. As researchers or practitioners in HCI, we should base ourselves on the present, look to the future, bravely face challenges, seize opportunities, continuously innovate, and promote the continuous advancement of this dynamic and promising field of HCI. Only in this way can we better utilize HCI technology, let technology truly serve people, and benefit human society.

The Interaction between Digital Transformation and HCI

In the wave of digital transformation, HCI plays a crucial role. As a bridge connecting technology and users, the development of HCI not only drives digital transformation but also is deeply influenced by it. From the perspective of ANT, digital transformation and HCI constitute a complex network composed of various actors, including technology, users, organizations, and social environments. In this network, each actor interacts and influences each other continuously, shaping the trajectory of digital transformation and HCI development collectively.

On the one hand, digital transformation provides new opportunities and challenges for HCI development. The emergence of emerging technologies such as AI, virtual reality/augmented reality (AR), and the IoT has opened up new possibilities for HCI design. These technologies not only change the way people interact with digital systems but also challenge traditional HCI design principles and methods. For instance, the development of AI technology enables more natural and intelligent HCI, but it also raises new issues regarding privacy, security, and ethics. On the other hand, the development of HCI also provides essential support and guidance for digital transformation. Well-designed HCI systems can significantly enhance the effectiveness of digital transformation, promote the adoption and application of new technologies. The research results in the HCI field, such as user experience design, ergonomics, and human-computer collaboration, provide important theoretical guidance and methodological support for the practice of digital transformation. For example, user-centered design concepts help ensure that digital transformation initiatives truly meet user needs, thereby enhancing user acceptance and participation. The interaction between digital transformation and HCI is fully reflected in various industries. In finance, retail, healthcare, education, manufacturing, and other fields, digital transformation initiatives have introduced cutting-edge HCI technologies and design concepts, bringing innovation to business models and user experiences. Conversely, the digital transformation practices in these industries also provide rich application scenarios and empirical cases for HCI research, promoting further development of HCI theories and methods.

Successful HCI cases, such as intelligent assistants and wearable devices, demonstrate the interaction between technology, users, and social environments in the process of digital transformation, driving the evolution of HCI design. Take intelligent assistants as an example; their design not only reflects the progress of AI technology but also embodies the changes in user needs and society's expectations for convenience and intelligent services. In various industries' digital transformation processes, HCI plays a key role. For example, in the financial industry, the application of virtual customer service representatives and chatbots is becoming increasingly popular. These AI-based systems can provide 24/7 services, answer customer inquiries, process transactions, and even provide personalized financial advice. For instance, Bank of America's "Erica" virtual assistant uses AI and predictive analytics to help customers manage their finances (Bank of America, n.d.). In the retail industry, AR technology brings more interaction and fun to the shopping experience. IKEA's "IKEA Place" app allows users to virtually place furniture in their homes, helping them make wiser purchasing decisions while reducing returns due to size or style mismatch (IKEA, 2017). In the healthcare industry, wearable technology benefits from the push of digital transformation. Devices like Fitbit and Apple Watch can monitor users' health indicators in real-time, provide personalized health reports and advice through synchronization with smartphones, and even predict the risk of certain health issues (Patel et al., 2012). In the education field, VR and AR technologies are creating immersive learning experiences (Google, 2017a). Google Expeditions allow students and teachers to visit remote geographic locations or historical scenes through VR, providing a new interactive and participatory learning mode. In manufacturing, collaborative robots (Cobots) play an important role in the digital transformation of the manufacturing industry. These robots are designed to work safely with human workers, performing tasks such as assembly, polishing, and painting. Companies like General Electric and BMW have successfully deployed Cobots on their production lines, improving flexibility and efficiency (Liu et al., 2024).

With the continuous deepening of digital transformation, HCI plays an important role not only in the above-mentioned industries but also in a wider range of fields, and the boundaries of HCI are constantly expanding. The emergence of the "More-than-human HCI" concept marks the transition of HCI from a human-centered approach to a more ecological and diversified development stage. This trend is highly consistent with the sustainable development goals of digital transformation, emphasizing the full consideration of the interests and well-being of non-human actors (such as animals, plants, and the environment) in the digital transformation process, promoting harmonious coexistence between humans and nature. The interaction between digital transformation and HCI is complex and close. As an important part of digital transformation, the development of HCI is not only driven and influenced by digital transformation but also provides crucial support for the successful implementation of digital transformation. From the perspective of ANT, examining the interaction between the two, we can more comprehensively grasp the opportunities and challenges of digital transformation and explore a more inclusive and sustainable development path. With the emergence of new concepts such as "More-than-human HCI," the interaction between digital transformation and HCI will enter a new stage that is more ecological, intelligent, and human-centric, contributing wisdom and strength to the sustainable development of human society and the Earth's home.

Technological Innovation and User Experience

In the wave of digital transformation, the field of HCI is undergoing unprecedented changes. Technological innovation and user experience, as two core elements of HCI development, are increasingly receiving attention from both academia and industry for their interaction and influence. From the perspective of ANT, the enhancement of technological innovation and user experience depends not only on the efforts of individual actors but also on the result of the collaborative actions of the entire actor-network.

Technological innovation serves as a significant engine driving HCI development. The emergence of new technologies such as AI, virtual reality/AR, and the IoT has opened up new possibilities for HCI design. Take haptic feedback technology as an example; its development has significantly improved the immersion and interactivity of virtual reality systems, providing users with a more realistic and natural experience (Pacchierotti et al., 2017). However, ANT theory reminds us that the success of technological innovation depends not only on the advancement of technology itself but also on its alignment with factors such as user needs and sociocultural environments. Only when new technologies can effectively meet users' real needs and are widely accepted and used in specific sociocultural contexts can their value be truly realized (Latour, 2005b).

User experience is a crucial indicator for measuring the success of HCI design. Digital transformation not only changes the way users interact with technology but also raises higher requirements for user experience. From the ANT perspective, the formation of user experience is not only influenced by technical factors but also closely related to users' cognition, emotions, and behaviors (Hassenzahl, 2013). To enhance user experience, HCI designers need to deeply understand users' needs and expectations and closely integrate them with the technological innovation process. Moreover, different sociocultural environments also have significant impacts on user experience. Cross-cultural studies indicate significant differences in user preferences and acceptance of interaction design across different cultural backgrounds (Reinecke & Bernstein, 2013). Therefore, HCI design needs to fully consider cultural diversity and provide user experience solutions that meet localization needs.

At the organizational level, digital transformation is reshaping the operation mode and value creation logic of traditional organizations. Traditional organizations are mostly centered around the physical world, emphasizing physical resources and offline operations, while digital organizations pay more attention to the development and utilization of the virtual world. They reconstruct and optimize business models, organizational structures, and value chains through digital technology and innovative thinking (Vial, 2021). This transformation is not only reflected in the digitization and automation of internal processes but also in the fundamental change in the interaction mode between organizations and the external environment. Digital organizations are more open, agile, and flexible, capable of rapidly sensing and responding to market changes, and collaborating deeply with customers, partners, and other stakeholders through digital platforms and ecosystems (Vial, 2021).

In the production field, digital transformation is driving the transition from traditional manufacturing to smart manufacturing and personalized customization. Traditional production models primarily rely on the physical world, relying on large-scale, standardized production lines to achieve economies of scale. Digital production fully leverages the advantages of the virtual world, realizing intelligent design, simulation optimization, and rapid iteration of products through technologies such as digital twins and additive manufacturing (Tao et al., 2019). This production model is more flexible and efficient, capable of meeting the increasingly diverse and personalized needs of users. At the same time, the deep integration of the virtual world and the physical world also provides strong support for real-time monitoring, predictive maintenance, and other aspects of the production process, contributing to the improvement of production efficiency and product quality (Kusiak, 2018).

In summary, the interaction between technological innovation and user experience in digital transformation can be summarized in the following aspects (Table 1.2).

These transformations reflect the profound impact of digital transformation on the HCI field, while also indicating new directions and possibilities for HCI research and practice. As digital transformation continues to deepen, the synergistic development of technological innovation and user experience will become an important proposition in the HCI field.

TABLE 1.2
The Interaction between Technological Innovation and User Experience in Digital Transformation

Dimensions	Traditional Models	Digital Transformation
HCI Design	Technology-Centric	User-Centric, emphasizing alignment between technology and user needs, as well as societal and cultural contexts.
Organizational Operations	Physical World-Centric, emphasizing physical resources and offline operations.	Virtual World-Centric, restructuring business models and organizational structures through digital technologies and innovative thinking.
Production Models	Primarily relying on the physical world, leveraging large-scale, standardized production lines.	Primarily leveraging the virtual world, implementing intelligent design and personalized customization through digital twins, additive manufacturing, etc.

When exploring the impact of digital transformation on HCI, it may be beneficial to approach it from a scenario construction perspective. Analyzing the impact of digital transformation on HCI can help us to comprehensively and systematically understand this complex and dynamic process (Bishop et al., 2007). By identifying key uncertainties and exploring multiple possible future scenarios, decision-making and strategic planning can be facilitated. In the context of digital transformation, the dimensions of physical world/virtual world and traditional/digital represent crucial uncertainties shaping the future development of HCI. By constructing a 2×2 matrix based on these dimensions and utilizing relevant future research methods such as Futures Triangle (Inayatullah, 2008), Futures Signals (Hiltunen, 2008), Futures Signs (Saritas & Smith, 2011), and Futures Wheel (Glenn, 2009), a more comprehensive analysis of the impact of digital transformation on HCI can be achieved. Additionally, employing the causal layered analysis (CLA) method (Inayatullah, 1998), allows for the exploration of the deeper factors influencing how digital transformation affects HCI, such as worldviews, values, and institutions. From the perspective of the experience triangle, the future development of HCI needs to comprehensively consider the dimensions of product, audience, and context (Hassenzahl, 2018).

On the other hand, analyzing the internal and external influences of human-driven and technology-driven factors is crucial for understanding the motives, processes, and outcomes of digital transformation. Human-driven factors emphasize a people-centric approach, focusing on the impact of digital transformation on individuals, organizations, and society, and emphasizing the harmony between technological development and human values. Technology-driven factors, on the other hand, focus on driving business growth and efficiency improvement through technological innovation, emphasizing the leading role of technology in digital transformation (Vial, 2021). The impact of digital transformation on HCI can be examined from the perspectives of the Anthropocene and Metamodernism. In this context, human-driven approaches need to consider the ecological effects and social ethical impacts of technology, while technology-driven approaches need to balance technological humanistic connotations and social responsibilities, considering the shaping role of technology on deep-seated factors such as user emotions and identity. The concept of the Anthropocene emphasizes the profound impact of human activities on the Earth's ecosystem, prompting us to reflect on the relationship between technological development and sustainability (Latour, 2017). In this context, human-driven digital transformation needs to pay more attention to the ecological effects of technology, ensuring the coordination and unity of the digital process with environmental protection. At the same time, human-driven digital transformation also needs to consider the impact of technology on social ethics and human values, avoiding the negative externalities of technology. Metamodernism represents a cultural trend of reconstructing meaning and value after the deconstruction of postmodernity (Vermeulen & Van den Akker, 2010). In the context of metamodernism, technology-driven digital transformation not only seeks efficiency and functionality improvement but also considers the humanistic connotations and social responsibilities of technology. This implies that HCI design should not only meet users' functional needs but also consider the shaping role of technology on deep-seated factors such as user emotions and identity (Table 1.3).

TABLE 1.3
The Shaping Role of Underlying Factors

Analytical Dimensions	Internal Influences	External Influences
Human-driven	Emphasize the ecological impact of technology to ensure the coordination and harmony between digitalization processes and environmental conservation.	Pay attention to the impact of technology on social ethics and human values, avoiding negative externalities of technology.
Technology-driven	Pursue efficiency and functionality enhancement while also considering the humanistic implications and social responsibilities of technology.	Consider the role of technology in shaping deep-seated factors such as user emotions and identity.

DESIGN THINKING AND FUTURES THINKING: SHAPING THE DIGITAL FUTURE

Integration of Design Thinking and Futures Thinking

In the digital age, the integration of design thinking and futures thinking has opened up new perspectives for HCI design. Design thinking emphasizes innovative approaches that are human-centered, technically feasible, and commercially viable (Brown, 2009), focusing on the process of innovative problem-solving and employing user-centered design methods and iterative experimentation. Futures thinking, on the other hand, focuses on exploring and predicting possible future scenarios, forming forward-looking innovative solutions based on this exploration. As mentioned earlier, the six pillars of futures thinking—mapping, predicting, timing, deepening, creating alternative futures, and transforming—provide HCI designers with a methodology to consider long-term social trends and technological developments during the design process, ensuring that design outcomes meet the needs and expectations of future users (Inayatullah, 2008).

The integration of design thinking and futures thinking not only enhances the foresight and capability to address complex problems but also encourages interdisciplinary collaboration. Sanders and Stappers emphasized the importance of co-creation, where designers, researchers, users, and other stakeholders participate in the design process, exploring and defining possible future scenarios and solutions through the integration of different perspectives and knowledge (Sanders & Stappers, 2014). Dubberly and Pangaro demonstrated how systems thinking can be applied in the design process, emphasizing the importance of considering interactions between systems when designing complex interactive systems (Dubberly & Pangaro, 2009). By combining systems thinking with design thinking and futures thinking, HCI designers can gain a more comprehensive understanding and design HCI systems that meet future user needs and technological trends.

Impact of Integrated Thinking on HCI

The combination of design thinking and futures thinking has profound implications for the future development of HCI. It encourages designers to not only focus on the application of current technology and immediate user needs but also anticipate the impact of future technological changes and societal shifts on user experience. This approach leads to a series of innovative HCI patterns, such as more natural language interaction, more efficient user interface design, and more personalized services. Furthermore, this integration also encourages interdisciplinary collaboration, allowing knowledge and theories from fields such as psychology, cognitive science, and sociology to be incorporated into HCI design, resulting in more comprehensive and in-depth design solutions.

As design thinking and futures thinking become more deeply integrated, the design principles of HCI are evolving. Designers are increasingly concerned with how to use technological innovation to address social issues, how to consider sustainability in design, and how to ensure that technological progress benefits all demographics. This evolution not only drives technological and methodological innovation in the HCI field but also promotes a sense of social responsibility and ethical awareness in design practice. By applying design thinking and futures thinking to HCI design, we can better understand the interactions between humans, technology, and society, contributing to the creation of a more humane, inclusive, and sustainable digital future.

Innovative Case Analysis

With the widespread application of design thinking and futures thinking in the HCI field, we have seen many impressive innovative cases. These cases not only showcase the latest advances in current technology but, more importantly, demonstrate how designers shape the future through forward-thinking and user-centered design methods.

Smart home control systems like Google Nest represent the application of design thinking and futures thinking in daily life (Google, 2020). By learning user preferences and automatically adjusting the home environment, these systems provide a more personalized and intelligent living experience. This design not only considers the current needs of users but also anticipates future expectations for home automation and energy efficiency. Similarly, wearable devices like Fitbit and Apple Watch demonstrate how design thinking and futures thinking are applied in the field of health management (Patel et al., 2015). These devices help users develop healthy habits through continuous monitoring and personalized feedback, reflecting designers' considerations for future health management models. In the customer service field, Alibaba's "Xiaomi" robot represents the innovative application of AI technology in HCI (Zeng, 2018). Through natural language processing and ML, "Xiaomi" can understand customer intentions and provide precise answers, greatly improving service efficiency and user satisfaction. This case demonstrates the development trend of future HCI, namely, using AI technology to create more natural and efficient interaction experiences. The application of virtual reality technology in education and training fields, such as Google Expeditions, demonstrates the potential of design thinking and futures thinking in reshaping learning methods (Google, 2017b). Through immersive experiences, students can learn knowledge in a more intuitive and interactive way, improving learning interest and effectiveness. This innovation not only addresses current pain points in education but is also based on thinking and exploration of future learning models. In the transportation field, Tesla's Autopilot autonomous driving technology demonstrates the application of design thinking and futures thinking in addressing urban traffic issues. By integrating advanced sensors and AI analysis, this technology can optimize traffic flow, improve road safety, and provide ideas for the construction of future smart cities (Tesla, Inc., n.d.).

These innovative cases demonstrate the application of design thinking and futures thinking in various fields and their tremendous potential in shaping the future of HCI. By being user-centered and based on insights into future technological and social trends, designers can create more intelligent, efficient, and personalized interaction experiences, bringing substantial improvements to people's lives. At the same time, these cases also inspire us that design thinking and futures thinking are not just methodologies but also a shift in mindset. It requires designers to break free from current limitations, think about future possibilities with an open and innovative mindset, and use interdisciplinary knowledge and skills to turn visions into reality. Only by embracing change and designing the future with foresight can we create HCI experiences that truly meet people's needs and drive social progress.

THE FUTURE OF HUMAN-COMPUTER INTERACTION: THE ROLE OF ARTIFICIAL INTELLIGENCE

The Impact of Artificial Intelligence on HCI

The rapid advancement of AI technology is profoundly shaping the future of the HCI field, becoming a key driver for innovation in interaction design. The introduction of AI not only enhances the intelligence level of devices but also enhances the personalization and adaptability of user experiences, while opening up new interaction patterns and application scenarios. The progress in NLP technology enables voice assistants to more accurately understand and respond to user voice commands, making interactions more intuitive and natural. ML algorithms, on the other hand, can continuously learn from user behavior and preferences to provide more customized services and recommendations. The application of these AI technologies greatly improves HCI design, providing users with smarter, more efficient, and insightful experiences. However, we also need to recognize the challenges AI application in HCI faces. Amershi, Cakmak, Knox, and Kulesza pointed out the crucial importance of ensuring the usability and accessibility of AI systems, as it relates to users' ability to effectively understand and control these systems (Amershi et al., 2014). Shneiderman criticized the risks of excessive reliance on AI, particularly in neglecting user control and transparency. He advocates for the adoption of "Human-in-the-Loop" design principles to ensure that the application of AI technology does not undermine human decision-making autonomy and agency (Shneiderman, 2020). These viewpoints indicate that when integrating AI into HCI design, we must carefully weigh its pros and cons and prioritize human-centeredness to ensure that technology always serves human needs.

To better anticipate the future impact of AI on HCI, we need to gain insights into technological trends and user expectations. Horvitz explored the intersection of AI and HCI, particularly in the application of AI for predicting user intents and needs. He introduced the concept of "mixed heuristics" to guide AI system design for better supporting human decision-making processes and enhancing the efficiency and effectiveness of human-computer collaboration (Horvitz, 2007). Research by Wang, Yang, Abdul, and Lim showcased the potential applications of AI in affective computing and user emotion recognition (Wang et al., 2019). These advancements indicate that future AI technologies will be able to create more sensitive and empathetic interaction systems, significantly improving user experiences.

Applications of AI in HCI: Case Studies

The author believes that foreseeing the development trends of AI in the HCI field is crucial for designing future-oriented HCI systems. By understanding the potential and limitations of AI, grasping user expectations and values, we can better guide technological innovation to shape a more intelligent, humane, and inclusive digital future. This requires HCI researchers and practitioners to maintain a forward-thinking mindset, actively embrace change, and uphold human-centered principles to ensure that AI technology always serves the goal of improving human life and societal well-being. Therefore, from the perspective of technological progress, such as the continuous development and integration of technologies such as AI, robotics, and the IoT, will drive innovation and breakthroughs in these fields. Second, in terms of user demand, people's expectations for smarter, more personalized, and efficient services and experiences are constantly increasing, which will drive the deepening application of AI in HCI. Furthermore, contemporary societal challenges, such as population aging, environmental degradation, and resource scarcity, highlight the potential key role of AI technology in addressing these challenges. Finally, in terms of value orientation, people's values toward technology are changing, placing more emphasis on human-centeredness, inclusivity, and sustainability, which will influence the development direction of AI technology. By analyzing multiple factors such as technology, demand, society, and values, we can foresee that AI will trigger extensive and profound changes in the future HCI field, creating a more intelligent, humane, and sustainable HCI ecosystem (Table 1.4).

FUTURE OPPORTUNITIES AND CHALLENGES

With the rapid advancement of AI technology, its application in the HCI field brings numerous opportunities for future interaction design while also facing several challenges. On the one hand, the introduction of AI technology is expected to significantly enhance the intelligence level of HCI and improve user experiences by opening up new interaction modes and application scenarios. On the other hand, ensuring the human-centric nature of interaction design, safeguarding user privacy and security, and balancing human-machine decision-making authority pose

TABLE 1.4
Anticipating the Changes AI Will Trigger in the Future HCI Field

Domain	Concept	Present	Near Future (5–10 years)	Far Future (10–20 years or beyond)
Embodied Intelligence	Combining AI with robotics to achieve higher levels of autonomy and adaptability to the environment, facilitating human-robot collaboration.	Preliminary AI robotic applications, such as smart home robots, industrial robots, etc.	Highly autonomous robotic systems capable of flexibly adapting to complex environments and seamlessly collaborating with humans.	Anthropomorphic AI robotic assistants with human-like cognitive, interactive, and learning capabilities, becoming partners to humans.
Empathetic Agents	Emphasizing the connection between AI and human emotions, improving human-computer communication through accurate emotion recognition and response.	Initial applications of affective computing, such as sentiment analysis, facial recognition.	More advanced empathetic AI systems capable of understanding complex emotional states and providing personalized emotional support.	AI assistants with genuine empathy, able to engage in emotional communication like humans, becoming emotional supporters and listeners.
Collaborative Networks	Highlighting the integration of AI with social systems, such as smart cities and social governance, showcasing AI's potential in improving resource efficiency and social participation.	Preliminary smart city applications, such as traffic management, energy optimization.	More comprehensive social collaborative networks, utilizing AI to optimize resource allocation and promote public participation in decision-making.	Highly intelligent, adaptive social systems with AI and humans collaboratively governing to achieve sustainable development and social equity.
Symbiotic Ecosystems	Exploring the harmonious coexistence of AI and the natural environment, emphasizing AI's crucial role in environmental monitoring, ecological protection, and sustainable development.	Initial AI environmental applications, such as wildlife monitoring, pollution prediction.	Broader AI ecosystems capable of real-time monitoring of environmental health, optimizing resource utilization, and promoting harmony between humans and nature.	Highly intelligent Earth ecosystem management systems, with AI becoming a key force in maintaining Earth's health and protecting biodiversity.

higher demands for the future development of HCI. As pointed out by Shneiderman, overreliance on AI technology may neglect user control and transparency, thereby weakening human decision-making authority and autonomy (Shneiderman, 2020). To address this challenge, future HCI design needs to adopt the "Human-in-the-Loop" principle, ensuring that AI technology operates under human supervision and control rather than replacing human judgment. This requires HCI researchers and designers to fully consider human needs and values when developing AI-driven interaction systems and to design modes of human-machine collaboration that enhance human capabilities rather than replace them.

Additionally, the application of AI technology in HCI also faces ethical and sociocultural challenges. For example, AI systems may amplify inherent biases in data, leading to unfair or discriminatory interaction design (Springer & Whittaker, 2019). Therefore, HCI designers need to adopt a more comprehensive perspective, focusing not only on technological performance but also on ethical and moral standards and social impact. This requires interdisciplinary collaboration, incorporating knowledge and insights from fields such as ethics, sociology, and psychology to jointly promote the development of responsible and inclusive AI technology (Wang et al., 2019).

Digital transformation also has a significant impact on the future development of HCI. The progress of emerging technologies such as virtual/AR and wearable devices provides more possibilities for innovative interaction design (Ashtari et al., 2020; Khakurel et al., 2020). At the same time, digital transformation requires HCI to adopt a more user-centric design philosophy, focusing on end-to-end user experience and integrating usability with business objectives (Sheng et al., 2021). HCI professionals are expected to play a more critical role in this transformation, driving organizational cultural change, promoting interdisciplinary collaboration, and translating research findings into practical applications (Bannon et al., 2018).

In conclusion, AI and digital transformation bring both vast opportunities and formidable challenges to the future development of HCI. To address these challenges and seize opportunities, the HCI field needs to adopt a more comprehensive, multidisciplinary, and human-centric research and practice paradigm. This requires us to uphold the values of human-centeredness, inclusivity, and responsible innovation while embracing technological innovations. Through human-machine collaboration, we can continuously explore and optimize HCI, ultimately realizing the vision of technology benefiting humanity and enhancing social welfare.

CONCLUSION

This chapter delves into the close relationship between digital transformation and future HCI, focusing on the impact of design thinking, future thinking, and AI technology on HCI development. In today's rapidly evolving digital age, understanding and guiding the progress of HCI are not only crucial for improving user experience but also key drivers of social and technological innovation. This study employs methods such as literature review, case analysis, and ANT to explore how digital transformation shapes the evolution of future HCI. By analyzing HCI development trends and technological innovations in different industries, the research reveals that the integration of design thinking, future thinking, and AI technology is becoming the core driving force behind HCI field development. The findings confirm that through the application of these methodologies and technologies, we can effectively address the challenges brought by digital transformation and promote the design and implementation of more efficient and human-centered HCI systems.

The main contribution of this chapter lies in the systematic analysis of the roles of design thinking, future thinking, and AI technology in promoting HCI development. Especially through the perspective of ANT, this study reveals the interactions and influences of multiple actors such as technology, users, and social environment in HCI development. Additionally, through multiple industry cases, the chapter demonstrates the practical application of the aforementioned theories and technologies, further promoting HCI innovation and evolution. These findings not only enrich the theoretical foundation of digital transformation and HCI but also provide valuable guidance and inspiration for practitioners. The work of this chapter not only answers the questions raised in the introduction but also provides new perspectives and methods for understanding and guiding the future development of HCI in the digital age. By emphasizing the integration of design thinking, future thinking, and AI technology, this study is of great significance for improving user experience and promoting technological innovation while laying a solid theoretical and practical foundation for building a more interconnected, intelligent, and human-centric digital world. In the future, HCI researchers and practitioners should continue to uphold the principles of human-centeredness, inclusivity, and responsible innovation, delve deeper into the integration of design thinking, future thinking, and AI technology, and continuously explore and optimize HCI through human-machine collaboration to ultimately realize the vision of technology benefiting humanity and enhancing social welfare.

REFERENCES

Amershi, S., Cakmak, M., Knox, W. B., & Kulesza, T. (2014). Power to the people: The role of humans in interactive machine learning. *AI Magazine*, 35(4), 105–120.

Ashby, S., Hanna, J., Matos, S., Nash, C., & Faria, A. (2019, November). Fourth-wave HCI meets the 21st century manifesto. In Proceedings of the Halfway to the Future Symposium 2019 (pp. 1–11).

Ashtari, N., Bunt, A., McGrenere, J., Nebeling, M., & Chilana, P. K. (2020, April). Creating augmented and virtual reality applications: Current practices, challenges, and opportunities. In Proceedings of the 2020 CHI Conference on Human Factors in Computing Systems (pp. 1–13). https://www.mdpi.com/journal/BDCC/special_issues/Virt_Reality

Bank of America. (n.d.). Erica. Retrieved October 7, 2024, from https://promotions.bankofamerica.com/digitalbanking/mobilebanking/erica

Bannon, L., Bardzell, J., & Bødker, S. (2018). Reimagining participatory design. *Interactions*, 26(1), 26–32.

Bishop, P., Hines, A., & Collins, T. (2007). The current state of scenario development: An overview of techniques. *Foresight*, 9(1), 5–25.

Bødker, S. (2015). Third-wave HCI, 10 years later—Participation and sharing. *Interactions*, 22(5), 24–31.

Bonanomi, M. M., & Bonanomi, M. M. (2019). Paradigms, Perspectives, and Context of Change for the Digital Transformation of Multidisciplinary Design Firms. Digital Transformation of Multidisciplinary Design Firms: A Systematic Analysis-Based Methodology for Organizational Change Management, 25–33.

Borin, L., & Edlund, J. (2018). Language technology and 3rd wave HCI: Towards phatic communication and situated interaction. New Directions in Third Wave Human-Computer Interaction: Volume 1-Technologies, 251–264.

Brown, T. (2009). Change by Design: How design thinking transforms organizations and inspires innovation. [Kindle 2 version]. Retrieved from Amazon.com.

Çebi, F., & Gözlü, S. (2023). Editorial technology management in digital transformation era. *IEEE Transactions on Engineering Management*, 70(7), 2463–2464.

Cinque, T., & Vincent, J. B. (Eds.). (2022). *Materializing digital futures: Touch, movement, sound and vision*. USA: Bloomsbury Publishing.

Dove, G., Chen, S., Fries, D., Johnson, V., Mydlarz, C., Bello, J. P., & Nov, O. (2022). From environmental monitoring to mitigation action: Considerations, challenges, and opportunities for HCI. *Proceedings of the ACM on Human-Computer Interaction*, 6(CSCW2), 1–31.

Dubberly, H., & Pangaro, P. (2009). What is conversation? How can we design for effective conversation. *Interactions Magazine*, 16(4), 22–28.

Frauenberger, C. (2019). Entanglement HCI the next wave? ACM Transactions on Computer-Human Interaction (TOCHI), 27(1), 1–27.

Garg, P., Gupta, B., Chauhan, A. K., Sivarajah, U., Gupta, S., & Modgil, S. (2021). Measuring the perceived benefits of implementing blockchain technology in the banking sector. *Technological Forecasting and Social Change*, 163, 120407.

Glenn, J. C. (2009). *Futures research methodology: Version 3.0*. T. J. Gordon (Ed.). Washington, DC: Millennium Project.

Google. (2017a). Google Expeditions [Mobile application software]. Retrieved from App Store (or Google Play Store, depending on where the app is).

Google. (2020). Google Nest: Smart Home Products & Solutions [Mobile application software]. Retrieved from App Store (or Google Play Store, depending on where the app is available).

Haktanır, E., Kahraman, C., Şeker, Ş, & Doğan, O. (2022). Future of digital transformation. *Intelligent systems in digital transformation: Theory and applications* (pp. 611–638). Cham: Springer International Publishing.

Hanelt, A., Bohnsack, R., Marz, D., & Antunes Marante, C. (2021). A systematic review of the literature on digital transformation: Insights and implications for strategy and organizational change. *Journal of Management Studies*, 58(5), 1159–1197.

Hassenzahl, M. (2013). User experience and experience design. *The Encyclopedia of Human-Computer Interaction*, 2, 1–14.

Hassenzahl, M. (2018). The thing and I: understanding the relationship between user and product. Funology 2: from usability to enjoyment, 301–313.

Hiltunen, E. (2008). The future sign and its three dimensions. *Futures*, 40(3), 247–260.

Horvitz, E. J. (2007). Reflections on challenges and promises of mixed-initiative interaction. *AI Magazine*, 28(2), 3. https://doi.org/10.1609/aimag.v28i2.2036

IKEA. (2017). IKEA Place App [Mobile application software]. Retrieved from App Store (or Google Play Store, depending on where the app is available).

Inayatullah, S. (1998). Causal layered analysis: Poststructuralism as method. *Futures*, 30(8), 815–829.

Inayatullah, S. (2008). Six pillars: Futures thinking for transforming. *Foresight*, 10(1), 4–21.

Jones, L., Nousir, A., Everrett, T., & Nabil, S. (2023, April). Libraries of Things: Understanding the Challenges of Sharing Tangible Collections and the Opportunities for HCI. In Proceedings of the 2023 CHI Conference on Human Factors in Computing Systems (pp. 1–18).

Kaivo-oja, J., & Knudsen, M. S., & Lauraéus, T. (2022) Future avenues of digital transformation. Public Innovation and Digital Transformation, 165.

Kallinikos, J., Aaltonen, A., & Marton, A. (2013). The ambivalent ontology of digital artifacts. *MIS*, 37(2), 357–370.

Kaptelinin, V., & Nardi, B. (2012). *Activity theory in HCI: Fundamentals and reflections.* Morgan & Claypool Publishers.

Kersten-van Dijk, E. T., Westerink, J. H. D. M., Beute, F., & IJsselsteijn, W. A. (2017). Personal informatics, self-insight, and behavior change: A critical review of current literature. *Human–Computer Interaction*, 32(5-6), 268–296. https://doi.org/10.1080/07370024.2016.1276456

Khakurel, J., Porras, J., Melkas, H., & Fu, B. (2020). A comprehensive framework of usability issues related to the wearable devices. Convergence of ICT and smart devices for emerging applications, 21–66.

Kim, Y. C. (2018). Alibaba: Jack Ma's unique growth strategy and the future of its global development in the Chinese digital business industry.The digitization of business in China: exploring the transformation from manufacturing to a digital service hub, 219–247.

Kusiak, A. (2018). Smart manufacturing. *International Journal of Production Research*, 56(1-2), 508–517.

Latour, B. (2005a). *An introduction to actor-network-theory. Reassembling the social.* Oxford: Oxford University Press.

Latour, B. (2005b). *Reassembling the social: An introduction to actor-network-theory.* Oxford University Press.

Latour, B. (2007). *Reassembling the social: An introduction to actor-network-theory.* Oxford, UK: Oxford University Press.

Latour, B. (2017). *Facing Gaia: Eight lectures on the new climatic regime.* John Wiley & Sons.

Liu, L., Guo, F., Zou, Z., & Duffy, V. G. (2024). Application, development and future opportunities of collaborative robots (cobots) in manufacturing: A literature review. *International Journal of Human–Computer Interaction*, 40(4), 915–932.

Markus, M. L., & Rowe, F. (2023). The digital transformation conundrum: Labels, definitions, phenomena, and theories. *Journal of the Association for Information Systems*, 24(2), 328–335.

Mukhlisah, Fauziah. (2023). Examine the competencies for upskilling in VUCA era (volatility, uncertainty, complexity and ambiguity) in Indonesia. *KnE Social Sciences.* https://doi.org/10.18502/kss.v8i11.13550

Norman, D. (2013). *The design of everyday things: Revised and expanded edition.* Basic books.

Oulasvirta, A., & Hornbæk, K. (2016, May). HCI research as problem-solving. In Proceedings of the 2016 CHI Conference on Human Factors in Computing Systems (pp. 4956–4967).

Pacchierotti, C., Sinclair, S., Solazzi, M., Frisoli, A., Hayward, V., & Prattichizzo, D. (2017). Wearable haptic systems for the fingertip and the hand: Taxonomy, review, and perspectives. *IEEE Transactions on Haptics*, 10(4), 580–600.

Patel, M. S., Asch, D. A., & Volpp, K. G. (2015). Wearable devices as facilitators, not drivers, of health behavior change. *JAMA*, 313(5), 459–460.

Patel, S., Park, H., Bonato, P., Chan, L., & Rodgers, M. (2012). A review of wearable sensors and systems with application in rehabilitation. *Journal of Neuroengineering and Rehabilitation*, 9, 1–17.

Pinhanez, C. S. (2020, July). HCI research challenges for the next generation of conversational systems. In Proceedings of the 2nd Conference on Conversational User Interfaces (pp. 1–4).

Reinecke, K., & Bernstein, A. (2013). Knowing what a user likes: A design science approach to interfaces that automatically adapt to culture. *MIS Quarterly*, 37(2), 427–453.

Roldan, W., Gao, X., Hishikawa, A. M., Ku, T., Li, Z., Zhang, E., … & Yip, J. (2020, April). Opportunities and challenges in involving users in project-based HCI education. In Proceedings of the 2020 CHI Conference on Human Factors in Computing Systems (pp. 1–15).

Rydenfält, C., & Persson, J. (2020). The usability and digitalization of healthcare: Third-wave HCI meets first-wave challenges. *XRDS: Crossroads, The ACM Magazine for Students*, 26(3), 42–45.

Sanders, E. B. N., & Stappers, P. J. (2014). Probes, toolkits and prototypes: Three approaches to making in codesigning. *CoDesign*, 10(1), 5–14.

Saritas, O., & Smith, J. E. (2011). The big picture–trends, drivers, wild cards, discontinuities and weak signals. *Futures*, 43(3), 292–312.

Seaborn, K., & Fels, D. I. (2015). Gamification in theory and action: A survey. *International Journal of Human-Computer Studies*, 74, 14–31. https://doi.org/10.1016/j.ijhcs.2014.09.006

Sheng, J., Amankwah-Amoah, J., Khan, Z., & Wang, X. (2021). COVID-19 pandemic in the new era of big data analytics: Methodological innovations and future research directions. *British Journal of Management*, 32(4), 1164–1183.

Shneiderman, B. (2020). Human-centered artificial intelligence: Reliable, safe & trustworthy. *International Journal of Human–Computer Interaction*, 36(6), 495–504.

Smith, A. C., Thomas, E., Snoswell, C. L., Haydon, H., Mehrotra, A., Clemensen, J., & Caffery, L. J. (2020). Telehealth for global emergencies: Implications for coronavirus disease 2019 (COVID-19). *Journal of Telemedicine and Telecare*, 26(5), 309–313.

Springer, A., & Whittaker, S. (2019, March). Progressive disclosure: empirically motivated approaches to designing effective transparency. In Proceedings of the 24th International Conference on Intelligent User Interfaces (pp. 107–120).

Suchman, L. A. (1987). *Plans and situated actions: The problem of human-machine communication*. Cambridge: Cambridge University Press.

Tao, F., Sui, F., Liu, A., Qi, Q., Zhang, M., Song, B., … & Nee, A. Y. (2019). Digital twin-driven product design framework. *International Journal of Production Research*, 57(12), 3935–3953.

Tesla, Inc. (n.d.). Autopilot. October 7, 2024, from https://www.tesla.com/autopilot

Tratkowska, K. (2019). Digital transformation: Theoretical backgrounds of digital change. Management Sciences. *Nauki o Zarzadzaniu*, 24(4), 32–37.

Vermeulen, T., & Van Den Akker, R. (2010). Notes on metamodernism. *Journal of Aesthetics & Culture*, 2(1), 5677.

Vial, G. (2021). Understanding digital transformation: A review and a research agenda. Managing digital transformation, 13–66.

Wang, D., Yang, Q., Abdul, A., & Lim, B. Y. (2019, May). Designing theory-driven user-centric explainable AI. In Proceedings of the 2019 CHI Conference on Human Factors in Computing Systems (pp. 1–15).

Weger, K., & Yeazitzis, T. (2023, September). Conceptualizing a Socio-technical Model for Evaluating AI-driven Technology. In Proceedings of the Human Factors and Ergonomics Society Annual Meeting (Vol. 67, No. 1, pp. 1639–1644). Sage CA: Los Angeles, CA: SAGE Publications.

Weritz, P. (2022). Hey leaders, it's time to train the workforce: Critical skills in the digital workplace. *Administrative Sciences*, 12(3), 94.

Wikipedia contributors. (n.d.). Actor–network theory. In Wikipedia, The Free Encyclopedia. Retrieved March 10, 2024, from https://en.wikipedia.org/wiki/Actor%E2%80%93network_theory

Xu, X., Hou, G., & Wang, J. (2022). Research on digital transformation based on complex systems: Visualization of knowledge maps and construction of a theoretical framework. *Sustainability*, 14(5), 2683.

Zeng, M. (2018, September-October). Alibaba and the future of business. Harvard Business Review. Retrieved October 7, 2024, from https://hbr.org/2018/09/alibaba-and-the-future-of-business

Zuboff, S. (2019). *The Age of Surveillance Capitalism: The Fight for a Human Future at the New Frontier of Power.* New York: Public Affairs.

2 Vision of Digital Futures
Novel Interactions for Futures Scenarios: The D\Tank Case Studies

Anna Barbara and Venere Ferraro

BACKGROUND

Futures studies cannot fail to consider the cardinal interaction between humans and computers, but figuring out what the forms, timing, modes, and themes on which these interactions may occur is not yet fully defined (Appadurai, 2014).

The advent of Artificial Intelligence (AI) is now within reach. Speculation exists in every direction: probable, possible futures, but the role of some institutions is to work on desirable scenarios, alternative presents, and innovation trajectories to achieve them.

The tools provided by Futures Studies aim to develop the ability through scenario to see and prepare for the future, to interpret significant changes, whether technological, political, or social and anthropological (Poli, 2017).

In the conventional sense, scenarios are narratives that depict various potential developments in the external environment. Each scenario delves into the possibility of different conditions fostering and adopting innovation. Scenarios can be categorized as either exploratory or normative. Exploratory scenarios begin with past and present trends, envisioning a plausible future, while anticipatory or normative scenarios are constructed based on diverse visions of the future, which can be either desired or, conversely, feared (Mietzner & Reger, 2005).

However, the relationship with the future that is sought to be established becomes more strategic than the future itself. It is no longer understood as a forecast, but as that long-term thinking that helps project design thinking into the future to consider how the context, in its broadest sense, may evolve (Corà, Fazio, & Collura, 2023).

Designing scenarios through Futures Studies methodologies requires disciplinary strengthening and broadening, both in terms of the methods of Design Thinking and the point of observation, which remains extremely human-centered (Inayatullah, Bussey, & Milojevic, 2006).

In this regard, the D\Tank team conducted two simultaneous research, namely "Design for Digital Care" and "Digital for Museums," with the goal of developing future design scenarios that integrate technologies and prioritize a human-centered approach.

D\Tank is a think tank created by the Department of Design at Politecnico di Milano and aim to promote design as a tool to transform advanced research into

DOI: 10.1201/9781032693606-3

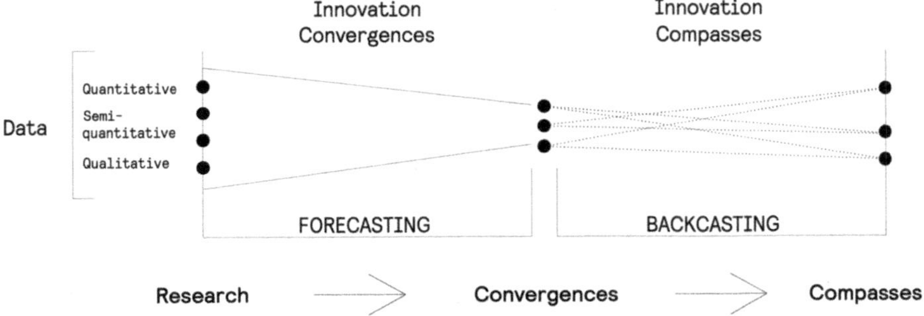

FIGURE 2.1 D\Tank approach and methodology.

innovation convergences for businesses and institutions. D\Tank uses a multiscale and multifocal approach to identify innovation trajectories through design practices and tools as a strategic lever to support the orientation in today's complexity.

Instead of focusing on "how the future will be like" D\Tank objective is to shape future technologies and re-design them to create "alternative present" (Auger, 2013). To envision such alterative presents D\Tank capitalizing quantitative, semi-quantitative, and qualitative methods likewise, literature review, patents analysis, Delphi method, horizon scanning and other methodologies and speculative practices (see Figure 2.1).

The Case Studies of Digital Care and Digital for Museum

The parallel research mentioned in the previous section engaged a total of 14 and 18 participants, including PhD students, research fellows, researchers, associates, and full professors. Several activities were undertaken to conceptualize these scenarios. Initially, a thorough examination of literature and patents was conducted to identify key challenges and themes in both digital care and digital museum domains. As a result, three primary clusters emerged within the digital care realm: Distributed Care, Self-Care, and Health Booster Technologies.

Distributed Care

Distributed Care involves leveraging digital technologies to optimize care delivery, creating a network that includes the community, caregivers, patients, and infrastructure. This concept introduces the digital caregiver as an intermediary, fostering humanized relationships and improving access to care. Central themes include community-centered care, physical-virtual communities, and the importance of digital literacy (Kringos, Boerma, Hutchinson, & Saltman, 2015; Milligan, 2016; Kor, Yanovsky, Pattinson, & Kharchenko, 2016; Johansson et al., 2021; Hallqvist, 2022).

On the other hand, Self-Care encompasses holistic processes related to hygiene, nutrition, lifestyle, and socioeconomic factors. Health applications, incorporating AI-based recommendation systems and integrated with wearable devices, play a pivotal role in disease prevention and the adoption of healthy lifestyles. Self-Care extends beyond disease prevention to include mental health and virtuous behaviors (Salim & Lim, 2019; Han, Ko, Jang, & Hwang, 2021).

Health Booster Technologies, particularly Mobile Health (mHealth) and the Internet of Things (IoT), empower individuals to monitor diseases, engage with healthcare providers, and adopt preventive behaviors, transforming healthcare to be more accessible and secure (Paglialonga et al., 2019).

Digital for Museum

Similarly, three main clusters were identified for the digital museum: Extended Rituals Experiences, Inclusive Engagement, and Museum-system management.

Extended Ritual Experiences represent a new paradigm in the museum, enhancing the integration of emerging technologies with exhibited heritage to increase visitor engagement through educational, emotional, and multisensory experiences (Lee, Jung, tom Dieck, & Chung, 2020; Batchelor, Schnabel, & Dudding, 2021).

Inclusive Engagement aims to make museum experiences accessible to a diverse range of visitors by incorporating certain practices and utilizing emerging technologies to enhance cultural accessibility. Participatory design approaches facilitate dialogue with communities and promote greater inclusivity (Cesário, Acedo, Nunes, & Nisi, 2022).

Regarding Museum-system management, the introduction of new technologies presents opportunities for utilization but also poses challenges related to intellectual property, regulation, and the digital transformation of organizational and relational skills within the museum structure (Taormina & Baraldi, 2022).

In conclusion, both digital care and digital museums are multifaceted and strategic topics intersecting advanced research, technological innovation, and a person-centered approach.

Design plays a crucial role in understanding users' needs, aspirations, and motivations.

Afterward, extensive user analysis was conducted in relation to identified macro-areas in both the Digital Care domain and Digital for Museums.

Queries through Google Analytics provided a lens for the analysis, and the identified topics were validated through purposeful focus groups with decision-makers in the care and museum fields.

The trajectories derived from literature review and patents analysis received strong approval and support from both users and experts.

The Near Future into Digital Care and Digital for Museums Domains

With these initial findings, the authors applied a typical approach, including back-casting and forecasting, to outline future design scenarios. Indeed, to produce systemic visions capable of creating convergences of innovation, D\Tank employs a multidisciplinary approach to scenario design.

It starts from the consideration that in the interaction between human and digital technologies, one cannot in fact think that the issue can be resolved solely on the technological level, but social spillovers, cultural matrices, political strategies, etc. must be considered. Through a multidisciplinary approach, the emphasis on

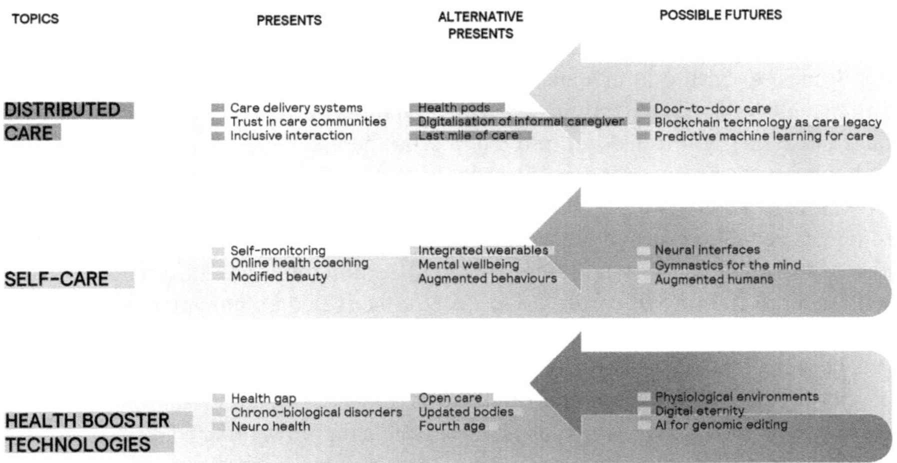

FIGURE 2.2 Digital care: possible futures and alternative presents.

predictions can thus be diminished in favor of greater multiculturalism and plurality (Sardar, 1999). By taking on the above-mentioned approach, the D\Tank, started from the present and envisioned possible futures and alternative presents both for Digital Care and Digital for Museum (see Figures 2.2 and 2.3).

The alternative scenarios related to Digital Care have been condensed into three categories: Last Mile for Healthcare, Digital Impatience, and Affordable Care.

In the first scenario, three main trajectories are emphasized: distributed care, healthcare delivery, and home care; they will enable several innovations. In the near future, digitization will facilitate the extension of care to cover the last mile,

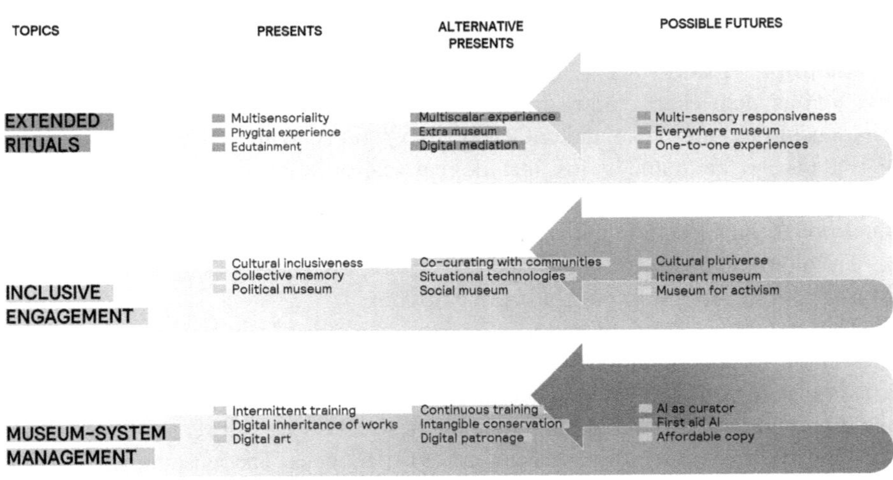

FIGURE 2.3 Digital for museum: possible futures and alternative presents.

reaching decentralized patients by providing care at home. Logistics, transportation, and sorting services will occur on dedicated infrastructures and fast means (including drones) to ensure immediate and widespread delivery. The healthcare network will establish distributed care nodes in the territory to reduce congestion in hospitals. The home will transform into a domestic therapeutic terminal, equipped with technologies to monitor, and connect the patient to the healthcare service network. Rooms will feature integrated sensors, transforming them into physiological environments capable of self-monitoring and managing therapies. However, digitization also necessitates capacity-building activities. Communities and informal caregivers will undergo training to ensure quality and effective communication with doctors, healthcare professionals, digital platforms, and patients.

The Impatient Scenario focuses on Tracking Technology, introducing the concept of the digital waiting room and digital coaching. Digitization enables patients to manage and book services in real-time, creating a digital waiting room that replaces the physical one, thereby changing waiting times, reducing anxiety, and minimizing delays. Apps, chatbots, and blockchain systems will manage appointments, organize care, and facilitate data sharing. Wearable devices for self-monitoring will connect the patient's body to diagnostic terminals through IoT and sensors. Access to real-time health information and the use of machine learning techniques will alter people's awareness and lead to augmented behaviors. Digital technologies will continue to support mental health, bridging physical and temporal distances between doctors and patients through real-time engagement via digital coaching, e-mental health technologies, and digital mental fitness services.

Affordable Care can be summarized in three key points: informal caregivers, digital literacy, and care pit stops. There is still a noticeable gap in access to healthcare delivery between people in different countries. Global health organizations propose addressing this issue through the democratization of healthcare services and equity in care. Creating interfaces using simplified languages and basic technologies will ensure inclusive interaction and literacy for everyone, including doctors, patients, healthcare staff, and caregivers. Overcoming the digital divide will guarantee accessibility and equitable service provision to every territory. This scenario envisions the creation of 24/7 care services offered by internet networks, enabling live and asynchronous communication between patients and doctors, making the relationship more flexible and detached from rigid schedules. The proliferation of hybrid places, facilitated by IoT, includes care capsules for well-being, equipped for quick diagnoses and small-scale treatments, distributed in offices, shopping centers, intermodal transportation hubs, and museums.

The alternative scenarios in Digital Care are represented by the following: Infinite Museum, À la carte Museum, and Next Museum.

The Infinite Museum scenario envisions a multiscale experience and digital mediation. In this scenario, the museum surpasses its traditional boundaries, with artworks dematerializing and collections being exhibited and narrated through new media supports. Extended Reality technologies will engage all senses, leading visitors toward more immersive and memorable experiences. The museum will become infinite, extending beyond its physical walls, and expanding into the network. Technologies such as AI will play a crucial role in delivering content in a versatile and engaging manner. This approach aims to make cultural heritage more accessible to diverse audiences across different locations and times. Edutainment will facilitate visitor interaction with artworks, ensuring an active and engaging visit. Data analysis, facilitated by machine learning systems, will personalize each visitor's experience by suggesting paths, selecting content, and enhancing the overall experience before and after the museum visit.

The À la carte Museum encompasses themes such as co-curating community, situational technologies, and the social museum. In this perspective, the museum will increasingly integrate with territories and communities. Curation will be a collaborative effort between experts and the community, with activities and programs forming a shared and inclusive project aimed at facilitating intercultural narratives and dialogues. Digital platforms will provide additional opportunities for relationships, interactions, and comparisons. Equity, accessibility, and inclusion will serve as crucial indicators, ensuring the engagement of visitors with physical and cognitive fragilities or belonging to cultural minorities. Extended Reality technologies will enable visitors to immerse themselves realistically in situational experiences, thereby reinforcing the connection with the place itself.

The museum will function as a cultural and democratic agora open to the visitor community. Both physical and digital spaces will be available to welcome proposals, debates, and experiments on social, political, and environmental themes, aligning with the nature of the place.

Last, the Next Museum envisions a future characterized by continuous training processes and immaterial conservation. Digitization is set to revolutionize museum management from within. The introduction of new professional skills will be achieved through ongoing and targeted staff training, alongside the recruitment of experts in emerging sectors. Digital literacy will become fundamental in proposing innovative museum forms and alternative modes of exhibition and interaction. Museum collections and archives will undergo digitization or be replicated through digital twins to ensure enhanced protection of material heritage. Conservation efforts will extend to both digitized and native digital artworks, necessitating the use of advanced technologies and nanomaterials to withstand the test of time and wear. Immovable heritage will also require conservation and active use, demanding an extension of the museum's and its experts' skills.

DISCUSSION

Extremely interesting questions emerged from D\tank's scenario building work on computer-human interaction, both on the scenarios themselves and their temporality, and on the method of constructing them (Amara, 1981).

The first issue concerns the timing of scenario projection. The issues decided by digital require scenarios of near futures, not extensions to 10–20 years as was the case in the past. Digital technologies accelerate changes, but do not necessarily accompany transformations that often have other timescales and speeds.

In human-computer interaction, it becomes clear that one issue concerns precisely the reaction time, involvement, assimilation that humans have with respect to the plethora of innovations and technologies available.

The timing of the projections that D\Tank has considered therefore is a compromise between the urgency of responses from digital technologies and on the other hand the timing of effective assimilation by people and companies.

Downstream of scholarly research on the issues of digital for curation and for museums, capable of having foundations rooted in the scientific disciplines of design the D\Tank has therefore worked methodologically through exercises of speculation on long timescales, and then back-casting with the construction of alternative presents capable of deciding by actions closer to the present what the range of choices might be for one or more of those desirable scenarios to come true (Miller, 2006).

Directly related to this issue is the consideration that at present the available digital technology is in fact hardly ever fully assimilated by potential users. The issue of the need for user literacy that makes the technologies that already exist accessible, understandable, and usable turns out to be an indispensable condition for a relationship with digital technologies to be possible. Indeed, in both research it was found that in the face of technological consistency there is not an equally well-established technological literacy on the part of museum and healthcare workers, visitors, or elderly patients.

In addition to the issue of the knowledge, skills and generation divide, there is also the issue of a still uneven and uneven transition. There are regions of the world and territorial folds, where the absence of an efficient network makes any proposal for innovation inadequate. While there is a concentration of services in some regions of the world, there are others where the infrastructure is either not present, inadequate, or outdated, defining a divide that is sometimes even more dramatic than the generational divide. Digital democratization is not fully accompanied by infrastructure equity making interaction problematic, dispersed, and ineffective (Sardar, 1999).

In this context, interactions need to be seen within a systemic, nonpoint framework that can ensure a homogeneous distribution from the source to the user, including the possibility of relocating services, extending them as far as they have ever been able to go, disseminating performance, and reaching the last mile to the end user without too many intermediaries, wherever he or she may be.

Devices and digital media have changed our relationship with physical space and people. They have changed the forms of time by favoring multi-temporality, simultaneity, etc., but above all they have manipulated the distances between things and people, redefining the organization of time in spaces, relationship, and processes (Floridi, 2015).

Digital changes the role of bodies and senses, which in proportion to the use of technologies, are challenged to compensate for immersive experiences, accessibility in augmented environments, and inclusiveness of frail users. Smart technologies reshape spaces, behaviors, and infrastructures according to needs, desires, and environmental conditions, and personalize the experience (Carpo, 2017).

Indeed, in both research studies it was found that the potential offered in personalization and customization of services is still partly to be explored. Footprint technologies, which require leaving a trace, are increasingly less invasive. They are portable objects used to shop, to navigate, to interface with urban environments, they are wearable computers that we carry with us without discomfort rather with awareness and desire for self-monitoring (Ling & Campbell, 2008).

Through the interaction between humans and computers, the possibility of self-tracking emerged for the first time, which means collecting data about oneself, analyzing it, and diagnosing it. This possibility is a strategic factor both on the level of awareness, on the level of health, and on the level of customization of services and experiences. Real-time monitoring is one of the most interesting frontiers within this interaction that is so close and so intimate between humans and computers (Light, 2006).

The revolution introduced by smart technologies has introduced new temporalities in the interaction between people and services through new forms of mediation, between humans, objects, and spaces, which these digital devices play out in acceleration, compression, dilation and simultaneity between needs and performance. Indeed, the digital can accommodate ever-changing functional, experiential, and emotional needs (Mohamed Hassanein, 2017).

CONCLUSION

This chapter emphasizes the role of Futures Studies in understanding the interaction between humans and computers. While the forms and dynamics of these interactions are not fully defined, the study aimed to work on desirable scenarios, alternative presents, and innovation trajectories related to the advent of disruptive technologies.

To this end, D\Tank, conducted research projects on "Design for Digital Care" and "Digital for Museums" employing a multidisciplinary and human-centered approach, aiming to shape future technologies into "alternative presents."

The research involved case studies in Digital Care and Digital for Museums, identifying clusters such as Distributed Care, Self-Care, Health Booster Technologies, Extended Ritual Experiences, Inclusive Engagement, and Museum-system Management.

Digital technologies have transformed relationships, spaces, and behaviors. They have manipulated distances, redefined the organization of time, and challenged bodies and senses. The generated scenario emphasizes the role of smart technologies in personalization, customization, and self-tracking. The introduction of smart technologies has introduced new temporalities in the interaction between people and services, accommodating ever-changing needs and creating new forms of mediation.

Through the two case studies authors wanted to highlight the strategic and multidisciplinary nature of Futures Studies, the innovative approaches of D\Tank, and the transformative impact of digital technologies on various domains, including healthcare and museums. Challenges related to timing, user literacy, and infrastructure disparities are acknowledged, emphasizing the need for a comprehensive and inclusive approach to shape the future.

ACKNOWLEDGMENT

Authors would like to express gratitude to the team of researchers who participated in both the research projects Digital Care and Digital for Museums by D\Tank.

REFERENCES

Amara, R. (1981). The futures field: Searching for dimensions and boundaries. *The Futurist, 15*, 1.

Appadurai, A. (2014). Il futuro come fatto culturale: saggi sulla condizione globale. Raffaello Cortina. (The Future as Cultural Fact. Essays on the Global Condition, Cortina, Milano).

Auger, J. (2013). Speculative design: Crafting the speculation. *Digital Creativity, 24*(1), 11–35.

Batchelor, D., Schnabel, M. A., & Dudding, M. (2021). Smart heritage – Defining the discourse. *Heritage, 4*(2), 1005–1015. http://doi.org/10.3390/heritage4020055

Carpo, M. (2017). *The second digital turn: Design beyond intelligence.* MIT Press.

Cesário, V., Acedo, A., Nunes, N., & Nisi, V. (2022). Promoting Social Inclusion around Cultural Heritage Through Collaborative Digital Storytelling. In International Conference on ArtsIT, Interactivity and Game Creation (pp. 248–260). Springer International Publishing.

Corà, T., Fazio, L., & Collura, F. (2023). Futures by Design. Progettare innovazione nella complessità (pp. 105–112). Guerini Next, Milano.

Floridi, L. (2015). *The onlife manifesto: Being human in a hyperconnected era* (p. 264). Springer Nature.

Hallqvist, J. (2022). The making of a professional digital caregiver: Personalisation and friendliness as practices of humanisation. *Medical Humanities, 48*(3), 347–356.

Han, W. B., Ko, G. J., Jang, T. M., & Hwang, S. W. (2021). Materials, devices, and applications for wearable and implantable electronics. *ACS Applied Electronic Materials, 3*(2), 485–503.

Inayatullah, S., Bussey, M., & Milojevic, I. (2006). *Neohumanist educational futures: Liberating the pedagogical intellect.* Tamkang University Press.

Johansson, V., Islind, A. S., Lindroth, T., Angenete, E., & Gellerstedt, M. (2021). Online communities as a driver for patient empowerment: Systematic review. *Journal of Medical Internet Re-search, 23*(2), e19910.

Kor, A. L., Yanovsky, M., Pattinson, C., & Kharchenko, V. (2016, December). SMART-ITEM: IoT-enabled smart living. In 2016 Future Technologies Conference (FTC) (pp. 739–749). IEEE.

Kringos, D. S., Boerma, W. G., Hutchinson, A., & Saltman, R. B. (2015). Building Primary Care in a Changing Europe; WHO: The European Observatory on Health Systems and Policies: København, Denmark.

Lee, H., Jung, T. H., tom Dieck, M. C., & Chung, N. (2020). Experiencing immersive virtual reality in museums. *Information & Management, 57*(5), 103229.

Light, A. (2006). Adding method to meaning: A technique for exploring peoples' experience with technology. *Behaviour & Information Technology, 25*(2), 175–187.

Ling, R., & Campbell, S. W. (Eds.). (2008). *The reconstruction of space and time: Mobile communication practices.* Transaction Publishers.

Mietzner, D., & Reger, G. (2005). Advantages and disadvantages of scenario approaches for strategic foresight. *International Journal of Technology Intelligence and Planning, 1*(2), 220–239.

Miller, R. (2006, December). From trends to futures literacy. In Centre for Strategic Education. Seminar Series Paper. Melbourne.

Milligan, C. (2016). *There's no place like home: Place and care in an ageing society.* Routledge.

Mohamed Hassanein, H. (2017). Utilization of 'Multiple Kinetic Technology KT' in Interior Architecture Design as Concept of Futuristic Innovation. *ARChive, Forthcoming.*

Paglialonga, A., Patel, A. A., Pinto, E., Mugambi, D., & Keshavjee, K. (2019). The healthcare system perspective in mHealth. m_Health current and future applications, 127–142.

Poli, R. (2017). *Introduction to anticipation studies* (Vol. 1). Springer.
Salim, A., & Lim, S. (2019). Recent advances in non-invasive flexible and wearable wireless biosensors. *Biosensors and Bioelectronics*, *141*, 111422.
Sardar, Z. (Ed.). (1999). *Rescuing all our futures: The future of futures studies* (No. 30). Praeger Publishers.
Taormina, F., & Baraldi, S. B. (2022). Museums and digital technology: A literature review on organizational issues. *European Planning Studies*, *30*(9), 1676–1694.

Section II

Futures Literacy in HCI

3 Futures Thinking and Futures Literacy

Shams Hamid

"YES, YOU CAN IMAGINE"

She asked me to imagine. I said, "It's not easy."
She said, "No, dear, it is easy; you already do."
So, I imagined the past, and I envisioned the future.
I conjured chaos and dreamed of harmony.
I envisioned things that were not real,
places that existed only in my mind, and people I never met.
I pondered the act of imagining itself, wondering if I truly exist!
I envisioned worlds, landscapes, faces,
the past, the present, the future, chaos, harmony, and nothingness,
all unfolding endlessly, ever-changing before the canvas of my mind.
Yes, I've imagined it all, and so can you!

Imagination, as explored in the poetic introduction, is not just an artistic endeavor but a fundamental skill in futures thinking and Futures Literacy (FL). It enables us to visualize and prepare for potential technological and societal changes.

Building on the foundational concepts introduced in the abstract, this chapter explores further into the transformative potential of Futures Thinking and FL. We shall look at the application of these frameworks into human-computer interaction (HCI) such that we identify how these frameworks address the technological trends and the ethical integration of these trends into our society. Let us commence by gleaning the present condition and the challenges it presents, setting the stage for a detailed discussion on innovative methodologies.

NAVIGATING THE FUTURE: INTRODUCTION

The landscape of human existence in the future is marked by numerous unresolved challenges and unpredictable variables that shape the trajectory of civilization. Considering the intricate nature of this puzzle, we need to develop expertise and strategies that will help us to understand, foresee, and shape events/outcomes. Futures Thinking and FL become the ultimate resources of the task at hand, equipping humanity and the societies as a whole with the means to transcend uncertainty, imagine different realities, and endeavor for a fairer and sustainable world.

This chapter examines how Futures Thinking and FL are incorporated in HCI using systematic methodologies and strategic frameworks for transforming imagined future scenarios into technically and ethically viable innovations.

DOI: 10.1201/9781032693606-5

ECHOES OF TOMORROW: VISIONARY INSIGHTS

In the context of the future, it could be intriguing to look back and remember the visionaries of the past, especially Arthur C. Clarke, whose futuristic insights inspire us to push beyond the constraints of our present reality and explore the uncharted territories of the imagination. Arthur C. Clarke envisioned a future transformed by technology in ways that resonate deeply with our discussion of Futures Thinking and FL. Let's listen in.

Arthur C. Clarke's 1964 prediction about the future is a notable example of futures studies in popular culture. He envisioned a world where technological advancements would revolutionize human life, foreseeing a future where people wouldn't need to commute and could work or study from anywhere, anytime. Additionally, he speculated that the most intelligent creatures on earth would not be humans, but machines (BBC Archive, 1964). This aligns with the discussions in this chapter regarding the transformative potential of technology in reshaping HCIs.

As we confront the challenges and opportunities of an uncertain future, Clarke's words inspire us to push beyond the constraints of our present reality and explore the uncharted territories of the imagination. Arthur C. Clarke envisioned a future transformed by technology in ways that resonate deeply with our discussion of Futures Thinking and FL.

In the next part, we are going to explore the practical applications of Futures Thinking and FL in order to understand how governments and businesses are systematically adopting them.

FROM VISION TO REALITY: PRACTICAL APPLICATIONS

Clarke's imaginative proposals push us to establish the boundaries of the future. By turning from the broad idea we now move forward to the specific and simulated examples in the field of HCI. This sub-component investigates how foresight methods are being utilized by governments and corporations to transform innovative concepts to practical innovations.

FRAMEWORKS OF THE FUTURE: THINKING AND LITERACY

In the fast-paced world of technological evolution, the future is not merely an intellectual quest for knowledge but a decisive action for informed decision-making. In this chapter "Futures Thinking and FL," we explore both the realm of anticipation and foresight with HCI as part of the tapestry of our digital futurescapes.

Many governments use foresight frameworks and methodologies in many areas of policy-making. It has been developed as their practice by most countries in the Organisation for Economic Cooperation and Development (OECD), as well as developed countries and, more recently, by developing country governments. International Labour Organization (ILO) and UNICEF have tool kits. The UN Development Programme has pioneered a new way of looking ahead in public service. ADB, FAO of the UN, OECD, UNEP, and WB published full-fledged studies and op-eds as well.

Innovation labs and centers have been set up in many countries to develop innovative solutions to the policy and public sector stuckness problems. The labs give the government room for trial and error, innovative thinking, and future-proofing to avoid short-sighted policies and investments. They employ forecasting and foresighting tools to plot the changing landscape, understand uncertainties and drivers, and spot opportunities for growth, investment, and development.

The 2020 senior executives survey of major European and US companies by the Nuremberg Institute of Market Decisions (NIM) showed that at least over 90% of managers in organizations have to use foresight in one way or another. About 34% of firms noted that there was a separate business unit responsible for strategic foresight, while the others said that strategic foresight activities were part of other departments or managers' other assignments.

BEYOND BINARY: EMBRACING DIVERSE FUTURES

Have you noticed anytime you type "futures" the app warns of an "error," alerting you of a bug? Could this be due to the app's adherence to traditional, deterministic binary thinking, where only one predetermined future is considered correct? However, a shift toward rainbow thinking, which acknowledges the multitude of possibilities within the future, could offer a solution. Embracing the concept of "futures" rather than a singular future underscores the idea that the future is inherently unknowable and can only be imagined.

In the realm of FL and Futures Thinking opting for the term "futures" of "future" is purposeful and meaningful. It highlights the idea that there are multiple futures as opposed to just one fixed outcome. Embracing this diversity and unpredictability allows futurists to develop a nuanced way of envisioning and influencing what lies ahead.

DEFINITION OF KEY TERMS: FUTURES THINKING AND FL

At the center of our research is the concept of Futures Thinking—an intellectual framework beyond the traditional ways of seeing to help us look at the uncertain tomorrow. In the frame of HCI, Futures Thinking comes out as the engine of discovery, stimulating professionals to move beyond the restrictions of the now and toward the prospects of possible, plausible, probable, and preferred tomorrows.

FUTURES THINKING

Futures thinking (the theory and methods) and foresight (the practical application) constitute a set of instruments intended to help their users pinpoint emerging concerns, resolve uncertainties, describe multiple scenarios; forge a common vision for a desirable future through broad participation, design radical innovations, adopt sound policies and strategies (Futures Thinking in Asia and the Pacific, 2020).

- Conceptual Foundation: The Futures Thinking approach provides a solid, comprehensive conceptual framework that makes it possible for individuals and organizations to have a holistic, large-scale picture of what the future could be.

- Proactive Exploration: It facilitates proactive investigation of the possibilities rather than reactive forecasting, with the ultimate aim of shaping/determining the variety of desirable future situations.
- Methodologies and Tools: Exemplifies various methods including envisioning, environmental scanning, and Delphi technique, which aid in alternative futures assessment including analysis of outcomes of different scenarios.
- Foresight vs. Forecasting: The key difference between Foresight and forecasting is the open-ended participatory manner of visualizing possible futures compared to the deterministic projection of the future.

Futures Thinking, Foresight, and Forecasting

Futures Thinking and Foresight provide methods and practical applications to overcome possible challenges, and uncertainties, and realize opportunities. Therefore, these approaches involve the active engagement of multiple stakeholders to envision and create the desired futures. Forecasting that uses historical data and a closed universe as a basis for assumptions is remarkably different from Futures Thinking and Foresight which are more about imagination, participation, and preference for desirable outcomes. Foresight, a specific futures studies subdomain, deals with prolonged trends and scenarios, and its applications range across 5–20 years. It is a different process from strategic planning, which normally operates in a shorter timeframe.

Forecasting implies making a future prediction based on the preceding data or patterns, generally supposing that the current trends will follow suit without changing drastically. Prediction is very similar to forecasting, and it is a systematic approach of trying to list possible future events, in which the data and specific quantitative methods like statistical models are involved.

Predictions are rarely used in futures studies of the present time since they remove the agency of individuals and organizations. Prediction takes for granted a closed universe, an assumption that prediction itself does not influence the result (Futures Thinking in Asia and the Pacific, 2020).

Futures Literacy

Concerning FL, what we do not know, or at least do not think about very often or in much depth, are the answers to the questions: In many cases, a lot of people can't come up with the questions of "the future," and "the methodology we use to anticipate the future," given their lack of "future literacy." FL is a skill such as reading and writing, and just as with the development of other abilities FL can be learned and discovered using the learning process (UNESCO, 2018).

FL helps people understand why and how one uses the future to prepare, plan, and navigate the complexity and novelty of societies. The community and individuals can go through structured learning-by-doing activities known as FL Laboratories (FLLs) where they learn about the roots of what they imagine and can also empower them to zoom-wide their action (UNESCO & PMU, 2023).

UNESCO has championed FL since 2012: the kind of competency that allows individuals to better grasp the essence of the future and how it factors into what they think and do.

Understanding Futures Literacy

Instead of just looking into the future, future literacy involves the ability to see and use the future as a tool and a framework for the purposes of planning and preparation. This skillset is a must-have to navigate complex societies we live in and the latest technology.

- Skill Development: Like reading or writing, FL can be developed with the help of structured experiences such as FLLs, where FLL participants can jointly explore, resolve, and understand what is actually the future, and how their assumptions about future could affect their life.
- Applied Practice: Participants apply their FL Skill to tackle the prevailing challenges. This in turn increases their decision-making capacity and their strategic thinking.

THE ROLE OF FUTURES THINKING IN HCI RESEARCH AND DESIGN WORKS

HCI research and design cannot ignore Futures Thinking, it is not only a theoretical issue but a practical one. Fundamentally, HCI aims at defining the points where human behavior meets technology innovation, to create interactive digital experiences that are user-friendly and eco-friendly. However, the world is increasingly subject to unpredictability and complexity. Hence, the approaches to design and development used in the past fail to offer an answer in the current environments. Futures Thinking plays the role of a compass for the HCI practitioners navigating the unexplored waters of tech evolution.

Using the futurescape embracing speculative inquiry, and Foresight methodologies designers and researchers can have a nuanced understanding of the potential futures that await us. It facilitates them to acknowledge challenges, spot opportunities, and construct a roadmap toward better achievements. By embracing a future-oriented design, practitioners learn to challenge their present assumptions, rethink outdated traditions, and welcome the unfamiliar with curiosity and openness.

The emergence of Futures Thinking in the context of HCI marks the start of this transition from reactive to visionary design. It signals about the paradigmatic shift. The new mindsets, approaches to HCI research and methods require us to think outside the box as we now assume that uncertain, complex and unstable environment is characteristic of the social systems where we carry out our designs of the systems.

Speculative Futures: Designing Tomorrow

Speculative thinking is now a core component of HCI highlighted in recent studies; therefore, new methods such as causal layered analysis (CLA) and FL framework offered by UNESCO are critical. These systems are fundamental for framing a world where technology and human requisites interact in complicated and unexpected manners. Zhu, Chao, and Fu (2024) noted the multifaceted nature of speculative thinking with the symbiotic interplay of humans, technology, and institutions, which plays a crucial role in third-wave HCI research. According to Zhu et al. (2024), creative thinking is the cognitive process that deals with the examination of possibilities, potentialities, and different futures, and it is needed to develop methods to carefully navigate the speculative landscapes and responsibly design HCI.

Keys Element of the Futures Literacy in HCI

The field of FL in HCI consists of multiple skills that will assist practitioners in foreseeing, navigating, and choosing the paths that future-related discourse and endeavors will take.

Strategic Foresight: Strategic Foresight is how HCI professionals foresee upcoming trends, point out impending disruptions, and envisage alternatives which give them the capability to make sustainable developments in the technological field by considering societal and environmental impact proactively.

Critical Thinking: Critical thinking is a skill that lets people analyze preconceived beliefs, get to the essence of dominant narratives, and see what new technology offers in terms of user experience, social norms, cultural patterns, and environmental cost, which brings a more sophisticated and reflective understanding of HCI.

Creativity and Imagination: Creativity and imagination empower HCI practitioners to go beyond conventional thinking patterns, break the boundaries of implausibility, ponder the transformative possibilities that question the status quo, and build new inventive design paradigms.

Ethical Sensitivity: Ethical sensitivity involves the ability to consider and address the ethical implications of technologies, which may include security, safety, privacy, equality, accessibility, and sustainability, and then make decisions that are ethical and cognizant in the formulation and implementation of digital experiences.

FL involves the ability to understand, navigate, and shape the front lines of future-oriented discourse and behavior of users within HCI. FL is a skill like swimming the more you practice the more your skill improves. FL helps develop a critical, creative, and holistic mindset that addresses emerging patterns and uncertainty and encourages a proactive attitude to the designing and building of the future we desire.

Concerning the HCI domain, the importance of future literacy is difficult to dismiss because such literacy provides practitioners with existent intellectual capital and structures necessary for envisioning and answering the challenges of technological changes and societal problems. It helps to anticipate future scenarios and respond and adjust to new technological developments, and social and environmental changes.

Employing Foresight tools such as trend recognition, CLA, scenario planning, and speculative design, HCI specialists recognize weak signals, emerging trends, power hierarchies, disruptive forces, and anticipatory assumptions, which shape technological advancements. This enables HCI practitioners to minimize the possibilities of unintended consequences, ethical conflicts, and societal reactions that are often linked to the emergence of new technologies, and develop technology that makes societies more inclusive, equitable, and sustainable.

Contemplating the intricate interplay between human cognition and technological innovation, one cannot overlook the profound role of imagination. As we delve into the theoretical frameworks and practical applications of Futures Thinking in HCI, it is essential to pause and reflect on the creative potential of the human mind. The following poem, "Imagination," invites us to explore the enigmatic realm where perception meets creation, shedding light on the transformative power of imagination in shaping our understanding of the future.

"Imagination"

What lies beyond perception
is beyond perception,
But how does it become visible
Through writing?
What I cannot see,
What I cannot hear,
How does it manifest
in my creation?
Amidst a dense darkness
when I cannot see,
Still, by imagining you
Dawn emerges from the darkness.
The one I cannot meet,
The one I cannot touch,
But how does that one
Appear in my dreams?
What you could not hear,
What I couldn't say,
Yet, without hearing or saying,
it is comprehended.
Your happiness is yours alone,
My happiness is mine alone,
Yet when I see you sad
Tears well up in my eyes

Design Innovation, User Experience in HCI

FL is a source of inspiration in HCI design and it offers great opportunities for user's experiences providing a new outlook of possible and potential solutions. By taking a futures-oriented perspective and thinking ahead, HCI practitioners go beyond and move toward the future, imagining bolder, transformative futures that defy current norms and by so doing, initiate new innovative solutions.

Using speculative design, design fiction, and experiential prototyping, HCI experts experiment with different scenarios for the envisaged future, invent devices almost impossible to realize today, and involve stakeholders in creative dialogue about the principles related to emerging technologies. Utilizing the speculative design technique, designers foster creativity, experimentation, and collaboration. This in turn leads to the production of more imaginative experiences and innovations in interaction between humans and machines.

Additionally, HCI practitioners can collaborate with stakeholders through co-created preferred futures via participative design, which puts forth the power of users in technological innovation. Through cultivating a culture of joint production, co-design, and co-innovation, HCI practitioners guarantee the tech encounters' responsiveness to the diverse criteria, ambitions, expectations, and aims of users. This could bring about heightened adequacy, relevance, and utility of the HCI domain in a period of rapid technological change and intensive societal modification.

Having the ability of future literacy represents a transitional attitude allowing HCI experts to foretell and shape future experiences of digital communication with beguilement, imagination, comprehension, and compassion. Embracing a futures-oriented perspective, practitioners of HCI exceed the hindrance of the present moment and embark on a voyage of discoveries with the unraveling of the concealed variances to the creation of a more egalitarian, ethical, and sustainable digital destiny.

The construction of FL competencies in HCI could result from a multilateral approach integrating aptitudes and skills, educational styles and initiatives, interdisciplinary cooperation, and knowledge. Futures literate practitioners and learners of HCI can handle the intricacies of technological innovations and societal transformations wisely and intelligently, emerging as key agents toward charting the course for the future development of HCI that is edging us in the era of rapid changes and uncertainty.

CLA AND UNESCO'S FRAMEWORK: ADVANCED METHODS

Strategic Foresight, established as a key element for guiding strategies by government and corporate planners, is backed up by the application of powerful methodologies like CLA or UNESCO's FL. Let us move on to these paradigms, so that we can grasp what they are applying in the structured investigation of future circumstances.

CLA and UNESCO's FL offer frameworks that go beyond traditional forecasting, allowing for a multidimensional exploration of potential futures.

- The CLA and UNESCO's FL methodologies introduce fresh perspectives on the future of HCI.
- These methods offer a broad basis of applying various analytical approaches.
- The FL skills enable HCI practitioners to be ready for the unexpected and to foresee the trends in the future.

- The adoption of a FL attitude in HCI makes one more proactive and embraces change and innovation in the HCI research and design process.

We prioritize those frameworks that increase our knowledge and implement this approach in HCI as well. Amid a plethora of theories, learning tools, and methodologies like scenario planning, STEEPLE, the Futures Triangle, Design Futures, and many others, we highlight CLA and UNESCO's FL framework as the most powerful tools that deepen our anticipatory assumptions and narratives about the future.

CLA promotes a comprehensive consideration of the factors that underpin present events through an intensive chain of causes and effects that in turn reveal many underlying assumptions and worldviews of the future. This approach is very much in tune with HCI where a strong grasp of sociotechnical dynamics lays the foundation for creative, relevant, and sustainable solutions. Through this process, the HCI experts can explore and redefine the aspects of HCI ethically thus making the innovations every bit as thoughtful as they are inventive.

Likewise, UNESCO's FL framework has a structured way to shape the power of futures as a capacity. Understanding when and why to 'use the future' is crucial for HCI professionals navigating a rapidly evolving technological landscape. This literacy enhances the readiness and the ability to imagine and implement HCI solutions, which are adaptable for a variety of scenarios occurring in the future, beyond expectations.

Frameworks for Future Visioning in HCI

The development of frameworks including CLA and UNESCO's FL becomes a crucial area, not only because they can trace out the various possible futures, but also the ability to evaluate and shape these possible outcomes toward sustainable and ethical outcomes. Zhu et al. (2024) suggest that there is an underscored requirement for more methodical and rigorous ways to use various futures in HCI research (Zhu et al., 2024). This statement supports the chapter's emphasis on taking into account these frameworks which offer a structured way of looking at the changes occurring in societal and technical aspects. Through the entrenchment of these frameworks in HCI practices, we can generate a discipline that is not just reactive but also proactive in its strategic Foresight.

The decision to prioritize these two methods out of the various tools (Design Fiction, Backcasting, Prototyping Tools, and Expert Interviews) available for Futures Thinking is based on their immense capacity to inform a gripping and motivating understanding of the future. CLA and FL are not only used as a toolkit but also as a way to create a futures-oriented mindset among HCI designers – the key skillset for developers aiming to build useful and responsible human-centric technologies.

The integration of CLA and UNESCO's FL into HCI practices offers substantial benefits. These methodologies increase our knowledge spans and enhance our creativity and predictability landscape within the HCI sphere. CLA dives into the fundamental factors and scripts that sustain tech trends by discovering the hidden assumptions, and the mental models of this domain and deconstructing the state of affairs.

UNESCO's FL consists of a set of skills for professionals that enables them to rethink the possibilities of technological advancement and user interaction. By merging these approaches, HCI professionals can go beyond the traditional forecasting and problem-solving methods so that they can have a more detailed and visionary exploration of future technologies and their societal implications.

These methodologies offer the ability to take the elements apart and assemble them back together showing us how the society has changed. Therefore, these methodologies are very appropriate for the HCI professional who is required to design an adaptive, inclusive, and futuristic system.

Next, we learn about the way these methods among others, offer unique advantages for anticipating and molding the future in HCIs.

Introduction to Causal Layered Analysis (CLA)

The CLA framework proposed by social futurist Sohail Inayatullah helps uncover and explains the multiple layers of causation involved in social phenomena. The four-part structure of CLA is the prism through which to view the different aspects, starting from the societal impacts to the latest technological advancements.

The Litany: The list can be defined as the superficial examination that concentrates only on events and perceptible phenomena. It covers the current events, hot issues, and matters of the day. This layer handles the symptoms or features of the problems that are deeper than that. Let us say that the subject is technology. In this case, the litany layer could include talks about the smartphone phenomenon or the development of social media platforms.

The Systemic Causes: Digging deeper than the topmost strata, the fundamentals are the factors that shape the broad picture. This layer looks at the main systems, institutions, and policies that form the basis of the trends presented in this layer. It encompasses the broader societal, economic, and political environments which is what mold behavior and decision-making. As an example, systems causes in the domain of Info Tech might consist in the examination of regulative frameworks, economic incentives, and power dynamics within the industry.

The Worldview: The worldview level will get into the underlying cultural, values, and narrative beliefs that support the structures. It is rejecting the existing ways of thinking, ideologies, and theories which underpin social norms and conduct. The second layer focuses on the dominant narratives that develop among individuals and communities in their interpretation of the sense of reality and the processing of events. With regard to technology, the worldview level can involve investigating notions regarding progress, innovation, and the role of technology in influencing the human society.

The Myth: Finally, the myth layer is the spiritual essence of human consciousness manifesting the archetypal stories, metaphors, and symbolic meanings. It studies the collective myths, ceremonials, and symbols that reveal a person's character and destination. This stratum explores the instinctive factors that determine human behavior and activate long-term ideological stories. In the technology realm, the myth component could be explored through narratives of culture being appalled by machines, fear of technological dystopia, or aspiration of the utopian society.

Through investigating issues from these four dimensions, CLA provides a complete view of complicated events, demonstrating the intertwined relationship between the superficial manifestations and the deeper dynamics. This multidimensional approach creates room for different perspectives and possible strategic responses across domains including technology, policy, and social change.

CLA helps develop critical thinking skills by considering issues from different perspectives. This ability can be vital in the field of HCI, due to the integrated nature of human and technical factors.

Implementing CLA For HCI Futures

The CLA process will be divided into four distinct phases during the workshop, with each phase designed to provoke deeper thinking and uncover new insights:

The Litany (Surface Level):

Questions Raised:

What are the current visible trends and challenges in HCI?

What events or technologies are influencing HCI today?

Outcome:

1. Participants identify and list apparent trends and issues, setting the stage for a deeper exploration of underlying causes.

Social Causes and Structures (Systemic Level):

Questions Raised:

What political, economic, and technological structures are shaping these trends?

How do current HCI practices and industry standards influence these structures?

Outcome:

2. A clearer understanding of the systemic factors that support or hinder progress in HCI, enabling a move toward strategic structural changes.

Worldview (Worldview Level):

Questions Raised:

What are the predominant beliefs and ideologies that shape the HCI field?

How do societal values and cultural perceptions affect the design and adoption of HCI technologies?

Outcome:

3. Insight into the collective cultural and psychological drivers that impact HCI innovation and acceptance, helping to predict resistance or embrace of future trends.

Myths and Metaphors (Metaphorical Level):

Questions Raised:

What are the fundamental stories or metaphors that define HCI?

How do these narratives influence our expectations and ambitions for the future of HCI?

Outcome:

4. Participants redefine the narrative of HCI by exploring or creating new metaphors that align with an envisioned future, fostering a transformative perspective.

Potential CLA Workshop Outcomes

The workshop will focus on applying CLA so that participants can understand and envision the future of HCI. This approach will reveal the popular topics and look at the deeper systemic issues, cultural contexts, and metaphysical stories that will shape HCI over the next decade. After the workshop, participants will be in a position to make decisions using a multidimensional approach to future challenges and transformations in HCI, ready to implement changes that will consider existing realities as well as possible futures.

The deliberate style through which this is being done guarantees that the future planning of the HCI is not only reactive to emerging technologies but also proactively molds the course of the field in compliance with the deft societal directions and tales. This however shows the alignment to ensure that the HCI solutions are not only technically sound but also are in tandem with the culture and hence reach out to more people.

Applying UNESCO's Futures Literacy Framework for HCI Futures

UNESCO FL Framework serves as the core of the shared competencies we need to master to increase our perception of the future and how effectively we can act upon it *(UNESCO, 2023)*. This framework assumes that while the future is unforeseeable, the specific possibilities to probe it exist, and there are ways for us to reveal, reframe, and rethink the anticipations together with the assumptions. Implementing this framework generates insight about changing aspects of the future and it is transferred from people and groups to individuals helping them make better decisions in the present. The framework is structured around three key phases: Reveal, Reframe, and Rethink, the three components of the workshop that will build and extend our perspective about future conditions.

Application of Reveal, Reframe, and Rethink in the Workshop

Phase 1: Reveal

Objective: Identify and examine the implicit assumptions and expectations about the future of HCI.

Questions Raised:

What are the current trends that define the HCI field?

Which technologies or behaviors are we assuming will be important in 10 years?

How do these assumptions affect our planning and strategy?

Outcome:

Participants will uncover their subconscious biases and baseline expectations about the future. This revelation helps in recognizing how their current understanding shapes their Foresight and planning processes.

Phase 2: Reframe

Objective: Challenge the current assumptions and explore alternative ways of thinking about the HCI future.

Questions Raised:

If we remove current technological limitations, what could the future of HCI look like?

How would changing demographic, economic, or environmental factors redefine the needs and functions of HCI?

What happens if emerging technologies like AI and quantum computing evolve differently from expected paths?

Outcome:

This phase broadens the horizon of participants, encouraging them to think outside conventional wisdom and consider radically different futures. By doing so, they can create more robust strategies that are adaptable to unexpected changes.

Phase 3: Rethink

Objective: Develop new strategies based on a refined understanding of future possibilities.

Questions Raised:

Given these new scenarios, what innovative solutions can we develop to address future HCI needs?

How should our current projects and priorities change to align with these new understandings?

What skills and resources will we need to navigate this potential future?

Outcome:

In the final phase, participants will integrate their new knowledge into applicable strategies and unique ground-breaking technologies which were informed by an overall thorough and flexible understanding of future possibilities. Through this process, participants learn a set of universal response patterns and create a climate of lifelong learning and flexibility in their organizations.

With a solid understanding of CLA and FL under our belts, we are well-equipped to apply these frameworks to the dynamic field of HCI. These methodologies are not only theoretical constructs but are practical tools that have profound implications for design and innovation.

While we delve into the landscapes of imaginative futures and methodologies such as CLA and UNESCO's FL, it is a must to ensure that we discuss the ethics that these futures may bring with them. Ethical issues should not only play second fiddle but rather should be at the fulcrum base that formulates technologies we dreamt of and built to boost human dignity and societal well-being. As HCI specialists, addressing these concerns helps us to predict enthusiastically and resolve the possible impacts of emerging technologies on different populations and environments.

The following subsection is dedicated to the ethical principles that promise to serve as a compass when it comes to making decisions and taking actions which will in turn guarantee the development of HCI is not only advanced but just and responsible as well.

ETHICAL IMPERATIVES IN FUTURES-ORIENTED HCI DESIGN

The HCI technology development, is associated with moral dilemmas that must be attended. The problems that might come up are for example, those, which are related to privacy, data security, accessibility, and the opportunity for enhanced surveillance and control. The ethical frameworks we decide to use will determine the sort of technologies that are just for everyone and are used to serve the best interests of society at large. Now, we talk about the principles that should be adhered to by HCI practitioners during the creation of systems that promote human rights, include everybody, and prove to be sustainable.

Human Rights and Human Dignity (UNESCO)

Ensure that AI systems respect, protect, and promote human rights and fundamental freedoms.

Ensure that AI systems contribute to a more inclusive, sustainable, and peaceful world, avoiding biases that may exacerbate inequality or harm marginalized groups.

1. Reference: UNESCO's "Recommendation on the Ethics of Artificial Intelligence" adopted by all 193 Member States in 2021 highlights the importance of human oversight and fundamental principles such as transparency and fairness (UNESCO, 2021).

Transparency and Explainability (OECD)

Maintain transparency in AI operations and decisions, ensuring that they can be explained in comprehensible terms to users and stakeholders.

Implement mechanisms that allow for the tracing of decisions made with AI assistance to ensure accountability.

2. Reference: OECD AI Principles outline the need for AI systems to be transparent and explainable to foster trust and understanding (OECD, 2019).

Robustness, Security, and Safety (OECD)

Design AI systems to be secure and robust from inception through deployment, ensuring they are safe under all operating conditions.

Continuously monitor and update AI systems to respond to vulnerabilities and threats.

3. Reference: The OECD emphasizes the importance of robustness, security, and safety to prevent AI systems from causing unintended harm (OECD, 2019).

Accountability (UNESCO and OECD)

Ensure that there are mechanisms in place to hold designers and operators of AI systems accountable for the societal impact of their use.

Engage in constant evaluation of AI systems' social, economic, and legal impacts.

4. Reference: Both UNESCO and OECD highlight the necessity for accountability in AI implementations, with clear policies on liability and redress (OECD, 2019; UNESCO, 2021).

Inclusiveness and Social Impact (UNESCO)

Design and deploy AI systems that contribute positively to social welfare, enhancing accessibility and reducing discrimination.

Include diverse demographic groups in the development process to ensure that AI systems do not perpetuate social injustice.

5. Reference: UNESCO's framework calls for a focus on inclusiveness, ensuring that AI technologies are accessible to all and benefit humanity broadly (UNESCO, 2021).

Environmental and Ecosystem Impact (UNESCO)

Assess and mitigate the environmental impact of AI technologies, promoting sustainability and reducing ecological footprints.

Consider the lifecycle impacts of AI systems from production to disposal.

6. Reference: UNESCO stresses the importance of considering the environmental impacts of AI, and advocating for sustainable practices in AI development and deployment (UNESCO, 2021).

Ethical dilemmas we face today are not just about averting the bad scenarios, but the proactive model of shaping the future where technology is just a means to the end of favoring humanity as a whole and not destroy it. Having seen the different lessons from our encounters with Futures Thinking and literacy, we now know where to look at for the tools to succeed in a world whose future would contain great complexities and also a lot of opportunities.

ENVISIONING HCI'S FUTURE: INSIGHTS FROM YUVAL NOAH HARARI

Yuval Noah Harari in his book "Homo Deus: A Brief History of Tomorrow," explores future possibilities in human evolution envisioning a future dominated by AI and biotechnological advancement where natural HCI is completely transformed.

According to Harari, AI that deeply enhances human capacities will not only surpass human abilities but also will be successful in practically all domains such as decision-making, creativity, and problem-solving. It may result in the humans evolving into cyborgs or bionic men, mixing up real and artificial intelligence. Moreover, Harari talks about the consequences of this transition which includes the possibility of humanities being surpassed in the decision space as algorithms and AI consume more autonomy.

The admonition of Harari offers a plausible scenario in which the intelligence of the AI and biotechnological developments will dominate in the future. Harari introduces a visionary scenario in which the relationship between people and technological innovations acquires a transformative essence that is unique. It is just one of the many plausible, and possible futures, however, it is important to keep in mind that there might be more scenarios presenting a very different picture from the one Harari has suggested. Through being future-oriented we realize that the world can be complex with so many possibilities and the unpredictability that comes with it. The future is full of uncertainties, therefore, exploration and imagining of future scenarios, will be our lifeline to prepare us to face the uncertainties of tomorrow and to guide us to a future that accommodates all, is equitable, and is sustainable.

Harari's imaginative framework for an illuminative future becomes a perfect setting to stay informed and explore the possibilities of HCI. Scrutinizing tendencies and courses in AI and biotechnology help acquire important information with the potential to change the ground of HCI and redefine how humans interact with technology.

Showcasing two upcoming technologies, phages, and neuro microelectrodes, we take a deep dive into their possible use in bridging the gap between man and machine. To curb bacterial infections or to wear a new hat of brain-machine interface, these technologies effectively address the issue of HCI in different fields.

INNOVATIONS SHAPING TOMORROW'S HCI

Bacteriophages, highly prevalent in nature, play essential roles in bacterial ecosystems, animal and plant well-being, and ecological processes. Despite their seemingly simple nature as entities that replicate using bacterial hosts, bacteriophages have significant potential to impact various natural processes due to the critical role bacteria play in ecological balance. Exploration of phage therapy demonstrates the diverse applications of bacteriophages beyond conventional medical uses. Through the utilization of phages, practitioners can address bacterial infections, extend food preservation, and tackle environmental issues (García-Cruz et al., 2023).

These applications will be further examined alongside discussions on neural microelectrodes and fictional dialogues, illustrating how emerging technologies shape the future landscape of HCI.

In recent years, implantable neural microelectrodes have found extensive utility in fundamental neuroscience research and clinical applications, serving as neuroprostheses and treatments for various neurological conditions like depression, epilepsy, and Parkinson's disease. This suggests their potential for future use in closed-loop sensing systems, which could aid in the recovery of patients with neurological disorders. For instance, individuals with disabilities might control prosthetic limbs using their thoughts, or epilepsy patients could receive preemptive intervention before seizures.

Neural microelectrodes hold the potential to revolutionize brain-machine interfaces, facilitating direct communication between the brain and external devices. This advancement promises to enhance mobility, restore functionality, and enhance the quality of life for those affected by neurological ailments (Wang et al., 2023).

The integration of emerging technologies in HCI has a vast impact on everything from individual health and well-being to environmental sustainability and the future of work. These advancements encourage us to consider Futures Thinking and FL when planning HCI for the future. Through humanistic, transhumanistic, and posthumanistic philosophical vantage points we can overcome dilemmas and design a better and more decent world. Speaking about our objectives many times, we are willing to gain a really good and pleasant future, where humans and technology are embedded in each other and they improve each other and HCI as a whole.

FROM HUMANISM TO POSTHUMANISM: PHILOSOPHICAL PERSPECTIVES ON HCI

As it becomes evident, the role which technology takes in shaping the society is mainly defined by people's philosophical concepts of the world. Thus, we dig into humanism, transhumanism, and posthumanism—the philosophies that determine the philosophy future of HCI. In this way, entertaining different viewpoints helps us see the repercussions of our tech-related decisions and the world of the future that we seek.

Humanism, transhumanism, and posthumanism are philosophical approaches that characterize, different worldviews in which human beings and interaction between humans and technology are treated as a matter of principle. In a nutshell, HCI has been influenced by such notions and schools of thought to make us realize that humans have a lot to do with tech.

Humanism, which revolves around the idea of humankind as the center and highlights humans as what the HCI is all about, shows a human-centered design in HCI. It tells us about the importance of creating technologies that are only meant to update user's potential, provide them with autonomy, and also be ethically sound. Humanist precepts influence HCI professionals to mold user interfaces that are user intuitiveness, and accessibility and are aligned to ethical values that reflect human needs.

In contrast to transhumanism, which stands out for its advocacy as to how human abilities can be improved by using technology. Transhumanism, the concept that technology has or should have (if not already) a crucial role in the fundamental improvement of the human body, directly impacts the development of AR/VR solutions, wearable devices, and BCIs in HCI. Technologies capable of prolonging the human mind, brain as well as natural abilities, giving people totally different options for the manipulations with the digital environment and emotions.

Posthumanism is a notion that tries to transgress the traditional notions of individual human identity and agency by bringing to the forefront the difficulty of demarcating between humans and machines. When we speak about a posthumanist outlook in the HCI, it motivates us to develop innovative ideas about human-computer interfaces including embodied cognition, cyborg anthropology, and speculative design. Posthumanist HCI will endeavor to leap over the human-centric biases of thinking and will tend toward the more inclusive understanding of technology and the shaping of human experiences.

The bottom line is that humanism, transhumanism, and posthumanism express different perspectives that HCI experts can use for the formation and the design of interactive systems. By combining these philosophical perspectives with HCI research and practice we will be able to spark creativity, focus on user well-being, and generalize matters concerning the interrelation between humans and technology.

FICTIONAL DIALOGUE EXPLORING PHILOSOPHICAL PERSPECTIVES

Paul Jorion explores the evolving relationship between humans and technology in the context of HCI's future, addressing Humanism, Transhumanism, and Posthumanism to help HCI practitioners understand the importance of Philosophical Perspectives and their potential impact on the future of HCI." (Jorion, 2021).

The following fictional dialogue between Steve, a humanist; Nick, a transhumanist; and Donna, a posthumanist relating the repercussions of Futures Thinking and FL at HCI provides a thought-provoking exploration of philosophical perspectives on the future of HCI.

Round 1: Reveal

Panelists share how Futures Thinking and FL are useful in creating their desired future.

Moderator: Welcome, everyone, to our dialogue on the value of Futures Thinking and FL in HCI. Let's begin with Round 1: Reveal. Steven, how do you believe Futures Thinking and literacy contribute to shaping the future of HCI from a humanist perspective?

Steve (Humanist): Futures Thinking and literacy empower us to envision HCI designs that prioritize human well-being and fulfillment.

By understanding potential future scenarios, we can craft interfaces that enhance user experience and promote autonomy, ultimately leading to a more harmonious relationship between humans and technology.

Nick (Transhumanist): From a transhumanist standpoint, Futures Thinking allows us to anticipate technological advancements that can augment human abilities and transcend biological limitations. By embracing FL, we can guide the development of HCI toward creating transformative technologies that enhance human potential.

Donna (Posthumanist): Futures Thinking and literacy enable us to reimagine HCI beyond traditional human-centric approaches. By considering diverse perspectives and potential futures, we can design interfaces that blur the boundaries between humans and machines, fostering inclusive and equitable interactions in our increasingly technologized world.

Moderator: Thank you all for sharing your insights. Now, let's move on to Round 2: Reframe.

Round 2: Reframe

Panelists identify the underlying assumptions in their opposite camps

Moderator: In this round, I'd like each of you to identify the underlying assumptions in the perspectives of your fellow panelists.

Steve (Humanist): Nick, it seems your perspective assumes that technological advancement inherently leads to human progress, overlooking potential ethical and societal implications. Donna, your posthumanist stance appears to prioritize the integration of technology without fully considering the potential consequences for human identity and agency.

Nick (Transhumanist): Steven, your humanist perspective seems to emphasize human well-being at the expense of embracing radical technological advancements that could potentially enhance human capabilities. Donna, your posthumanist viewpoint may overlook the potential benefits of integrating technology into human experience, focusing too heavily on the preservation of human identity.

Donna (Posthumanist): Steven, your humanist perspective appears to center on the idea of human superiority and autonomy, potentially neglecting the agency of non-human entities and the interconnectedness of all beings. Nick, your transhumanist stance seems to prioritize technological solutions without fully considering the social and ethical implications of augmenting human abilities.

Moderator: Thank you for reframing each other's perspectives. Now, let's proceed to Round 3: Reframe continues.

Round 3: Reframe continues

Panelists reassess their underlying assumptions pointed out by others and explain their underlying assumptions.

Steve (Humanist): Upon reflection, I realize that my humanist perspective may indeed prioritize human autonomy and well-being without fully acknowledging the potential benefits of integrating technology into our lives. However, my underlying assumption remains rooted in the belief that human values should guide technological development to ensure it serves the greater good.

Nick (Transhumanist): I acknowledge that my transhumanist perspective may overlook the ethical and societal implications of radical technological advancements. However, my underlying assumption is that technology has the potential to enhance human capabilities and improve overall well-being, provided it is developed and implemented responsibly.

Donna (Posthumanist): While I understand the concerns raised about my posthumanist perspective, my underlying assumption is that embracing technology can lead to more inclusive and equitable interactions between humans and non-human entities. By blurring the boundaries between humans and machines, we can foster a deeper understanding of our interconnectedness and co-create more sustainable futures.

Moderator: Thank you for reassessing your assumptions and providing further insight. Now, let's move on to Round 4: Rethink.

Round 4: Rethink

Panelists appreciate each other's positions and the value of Futures Thinking and FL

Steve (Humanist): I appreciate the perspectives shared by both Nick and Donna. While we may approach HCI from different philosophical standpoints, it's clear that Futures Thinking and literacy play a crucial role in shaping our understanding of the evolving relationship between humans and technology.

Nick (Transhumanist): Likewise, I value the insights offered by Steven and Donna. Despite our differing views, it's evident that Futures Thinking and literacy provide us with the tools to navigate complex ethical and societal considerations in

HCI, ultimately guiding us toward more responsible and inclusive technological futures.

Donna (Posthumanist): I echo the sentiments expressed by Steven and Nick. Our dialogue has highlighted the importance of embracing diverse perspectives and considering the broader implications of technological development. Futures Thinking and literacy serve as invaluable resources for fostering dialogue and collaboration in shaping the future of HCI.

Moderator: Thank you all for your thoughtful reflections and contributions to this dialogue. The dialogue between Steve, Nick, and Donna illustrates the importance of Futures Thinking and FL in HCI. By revealing their perspectives, reframing underlying assumptions, reassessing these assumptions, and rethinking their positions, they demonstrate a collaborative approach to envisioning and shaping a future that is ethical, inclusive, and innovative. This dialogue serves as a model for HCI practitioners to integrate diverse philosophical perspectives into their work, ensuring technology's development aligns with human values and societal needs.

Steve (Humanist): Suppose for a moment we were talking about human-computer interaction, or HCI, as futures studies, or how to design interactions that meet our needs where we are today and not where we might be in the future. A very exciting enterprise I reckon! The forever dynamics of technology can raise some exciting trends that ought to be addressed by HCI role players too. And they can be equated to a live-philosophy playground.

Nick (Transhumanist): Ah, the future! As a transhumanism supporter, I strongly believe that we are very close to a technological innovation having a direct influence on our lives and hence an evolutionary development of our species. AI-related developments as well as human augmentation are flights into the unknown demand that we have FL and futures-oriented thinking.

Donna (Posthumanist): Oh, I cannot help but tell you that applying Futures Thinking to HCI is very close to my notion of posthumanism. This is no longer about trying to predict the future, rather it is about setting the course of what that future could become. We have to go through the technical factors and also the socio-economic, the political that are implied, and the multidimensional interactions among humans, machines, and the environment together.

Steve (Humanist): What you said is correct! To the advantage of humanhood, humanism stresses the mind and the reason of humans along with their ability to solve their problems and lead their lives. In the era of HCI, we need to be concerned that it will give humans more happiness and develop person-to-machine relationships, which are meaningful.

Nick (Transhumanist): Yet, you would not realize the capabilities if these did not take place. This will help us approach the unknown future and beyond. Through enhanced AI and brain-machine interfaces, we can get over the biological restrictions to ascend to worlds of sensibility new which were previously impossible to imagine.

Donna (Posthumanist): It is for free their curiosity, but as Nick, we must not overlook the ethical considerations. We are being called upon by posthumanism to critically examine the old world order which is based on social inequality and hence pave the way for just interactions in our evolving world.

Steve (Humanist): Well said, Donna. The more we focus on the distant future of HCI, we should not divert to our human traits amidst the buzz of innovativeness.

Nick (Transhumanist): Agreed! The essence of Futures Thinking is to make use of that capability so that we arrive at a world where both humans and technology are in symphony and the boundaries of humanity are stretched to greater levels.

Donna (Posthumanist): Hear, hear! As such, we can co-create a future that goes alongside diversity, strengthens the individual, and respects the range of possibilities of the people.

The dialogue between Steve, Nick, and Donna encapsulates the conversations about Futures Thinking and FL that hovers around HCI. Each contestant represents a separate philosophical school – Steve is the one that stands for humanism, Nick is for transhumanism, and Donna takes the side of posthumanism. With the adoption of humanism, transhumanism, or posthumanism, humans can refine their perceptions of what and what cannot be possible in the future for HCI. This raises crucial questions: What is the bearing of these philosophies on the concepts of technological innovation, ethical contemplations, and social implications? Are these traits going to affect the level of our expectations and preferences with regard to future developments of HCI?

INTEGRATING FUTURES THINKING AND FUTURES LITERACY INTO HCI FUTURES

The futures' thinking and Foresight techniques allow actors to handle emerging issues, unpredictability and opportunities by partnering with diverse stakeholders to craft and shape desired futures. Unlike traditional forecasting based on past data and assumes a closed universe, Futures Thinking and Foresight emphasize imagination, participation, and the focus on the desired outcomes. Foresight, the futures studies subfield, examines long-term trends and scenarios to inform strategic decisions, with applications spanning 5–20 years. It differs from strategic planning, which typically operates on shorter timeframes.

Integral to these processes is FL, a capability that enhances our understanding of the future's role in shaping current decisions and actions. As a foundational skill, the practice of FL enables people and organizations to discern and explore potential futures more consciously and creatively. Through the integration of diverse perspectives and aspirations, Futures Thinking, Foresight, and FL foster innovation and robust policy development to address complex societal challenges. By enhancing our capacity to navigate the unknown, FL ensures that our journey toward innovative and inclusive futures is both informed and intentional.

By embracing a futures-oriented mindset, HCI practitioners and scholars can ensure that they are not only reacting to changes but actively shaping the future of technology in ways that are ethical, inclusive, and forward-thinking. The insights gathered here should serve as a knowledge base and a call to action for continuous innovation and ethical vigilance. Looking ahead, the integration of interdisciplinary and transdisciplinary approaches will be crucial in navigating the uncharted territories of the future.

TOMORROW'S LESSONS: UNVEILING CHAPTER INSIGHTS

Ultimately, this chapter has shown us the powerful tools and strategies as being essential in creating a future HCI which emphasizes on a proactive and ethically grounded approach.

Imagination as a Tool: Unleashing the human imagination is indispensable for conceptualizing and convenience in HCIs.

Historical Insights: Contextualizing the journey of technology and its advancements by reviewing previous predictions, such as those from Arthur C. Clarke, allows humans to understand this path.

Methodological Focus: CLA and Future Literacy are the two key methods that one would use to strengthen futures' understanding.

Practical Applications: The Futures Thinking is not only a theoretical matter but it is in practice through policy-making and innovation in different sections of society.

Strategic Importance: By incorporating future literacy into the HCI practitioner's repertoire, responsive technology for upcoming societal requirements can be developed.

Diversity of Futures: Beyond predicting one singular "future", this point illustrates that a plurality of different "futures" is possible.

Ethical Considerations: Ethical Foresight in technological advancement is an integral part of HCI solutions that are inclusive and sustainable.

Interdisciplinary Relevance: The constantly changing nature of HCI makes the idea of Futures Thinking and literacy significant across different areas, helping us combine creativity and strategic planning.

As we conclude our exploration of FL in HCI, we are reminded of the timeless questions that have captivated human curiosity for centuries. The following poem, "Wonder," ponders the mysteries of existence and the interconnectedness of all things, and prompts us to contemplate the profound implications of Futures Thinking on our understanding of self, society, and the environment. Let us embark on this journey of wonder, where questions outnumber answers, and the pursuit of knowledge knows no bounds.

"WONDER"

I wonder and wonder what it is to be.
How am I, and how is everything?
How is everything connected with everything?
If everything is made of the same thing
How am I me and not everything?
A wonder that never ceases to be,
to be is to be a wonderful thing.
Oh, wonder! And wonder is a wonderful thing.
I wonder and wonder what it is not to be.
Will we have tomorrow, what we have today?
Why are you not here now when you were yesterday?
Where did you go? Where are you hidden?
Have you sacrificed yourself and become everything?
Or have you turned into nothing?
Are you, or are you not anymore?
Or, you are, and you are not as well?
You were connected to everything.
Yet, you were different from everything.
You were a wonder, a wonderful thing.
A wonder that never ceases to be,
to be, and not to be, is a wonderful thing.
Oh, wonder! And wonder is a wonderful thing.
I wonder and wonder how related are nothing and being.
If nothing comes from nothing, then how is there anything?
If everything comes from something, then how is there nothing?
Or nothing and being are entangled together?
A relationship that connects them and separates them
regardless of distance, regardless of time.
A wonder that never ceases to be.
The relationship between nothing and being is a wonderful thing.
Oh, wonder! And wonder is a wonderful thing.

REFERENCES

ADB (2020). Futures Thinking in Asia and the Pacific: Why Foresight Matters for Policy Makers. https://www.adb.org/sites/default/files/publication/579491/futures-thinking-asia-pacific-policy-makers.pdf

Relation to Chapter: Provides insights into how various governments within the Asia-Pacific region utilize Foresight and Futures Thinking in policy-making, relevant to the chapter's section on practical applications of these concepts in governance.

Arthur C. Clarke, writer, futurist, and inventor envisions the Future | Horizon | Past Predictions | BBC Archive. (1964). https://www.youtube.com/watch?v=YwELr8ir9qM&t=131s

García-Cruz, J. C., et al. (2023). Myriad applications of bacteriophages beyond phage therapy. *PeerJ*, 11, e15272. https://doi.org/10.7717/peerj.15272

Relation to Chapter: Cited to discuss the potential of bacteriophages in technology and HCI, relevant to the chapter's sections on biotechnological integration.

Harari, Y. N. (2016). *Homo Deus: A brief history of tomorrow.* Harvill Secker.

Relation to Chapter: Provides the futuristic context discussed in the concluding section of the chapter, especially regarding AI and biotechnology's impact on society.

Inayatullah, S. (2023) paper titled "Causal Layered Analysis: Poststructuralism as method" examines the application of poststructuralism in causal layered analysis. Metafuture: https://www.metafuture.org/2023/04/28/causal-layered-analysis-poststructuralism-as-method/

Relation to Chapter: This seminal paper by Sohail Inayatullah introduces CLA, which is utilized in the chapter to discuss how HCI can benefit from deeper Foresight methodologies.

Jorion, P. (2021). *Humanism and its discontents: The rise of transhumanism and posthumanism.* Palgrave Macmillan.

Relation to Chapter: Paul Jorion delves into the evolving relationship between humans and technology within the context of HCI's future through discussions on Humanism, Transhumanism, and Posthumanism relevant to the chapter's sections on Philosophical Perspectives.

Miller, R. (2021). Report of The Futures Literacy Lab-Novelty Online, Egyptian Youth Rethink the Future of Wellbeing in 2050. UNESCO's Library.

Relation to Chapter: This report offers a detailed example of a Futures Literacy Lab application, supporting the chapter's focus on UNESCO's methodologies for applying Futures Literacy.

OECD (2019) document on "AI Principles" outlines guidelines for ethical AI development and implementation. https://www.oecd.ai/en/ai-principles

Relation to Chapter: Cited in the ethical considerations section, outlining the principles for AI that promote transparency, accountability, and fairness, which are advocated for in the HCI practices discussed.

UNESCO (2018) "Transforming the Future: Anticipation in the 21st Century" explores the role of anticipation in shaping future outcomes. Paris: UNESCO. https://unesdoc.unesco.org/ark:/48223/pf0000264644

Relation to Chapter: This book, edited by Riel Miller, is one of the primary sources for the discussions on a central theme of this chapter Futures Literacy.

UNESCO. (2023). *Futures literacy & foresight: Using futures to prepare, plan, and innovate* (SHS/2023/PI/H/6). UNESCO. Retrieved from https://unesdoc.unesco.org/ark:/48223/pf0000386511

Relation to Chapter: This source elaborates on the theoretical and practical applications of Futures Literacy as advocated by UNESCO, aligning with the methodologies discussed in the chapter.

UNESCO (2021). Recommendation on the Ethics of Artificial Intelligence. https://www.unesco.org/en/articles/recommendation-ethics-artificial-intelligence

UNESCO, & Prince Mohammad bin Fahd University (Saudi Arabia), Center for Futuristic Studies. (2023). *Futures literacy laboratory playbook: An essential guide for co-designing a lab to explore how and why we anticipate.* UNESCO. https://doi.org/10.54678/KSWO4445

Relation to Chapter: This document is used to back the discussions on the ethical considerations necessary in the development of HCI technologies, as mentioned in the chapter.

Wang, Y., Yang, X., & Zhang, X., et al. (2023). Implantable intracortical microelectrodes: Reviewing the present with a focus on the future. *Microsyst Nanoeng*, 9, 7. https://doi.org/10.1038/s41378-022-00451-6

Relation to Chapter: Supports the chapter's exploration of emerging technologies such as neural microelectrodes and their implications for the future of HCI.

Zhu, L., Chao, C., & Fu, Z. (2024). How HCI integrates speculative thinking to envision futures. *Journal of Futures Studies*, 28(4). https://jfsdigital.org/how-hci-integrates-speculative-thinking-to-envision-futures/

Relation to Chapter: Directly correlates with the chapter's discussions on integrating speculative thinking within HCI, providing a case study and theoretical background.

I originally composed the poems "Yes, You Can Imagine," "Imagination," and "Wonder" in my native Urdu language. I translated them into English for their relevance to the theme of this chapter.

Relation to Chapter: Each poem celebrates the potency of imagination, creativity, critical thinking, empathy, intuition, and wonder aligning closely with themes such as Futures Thinking, Futures Literacy, HCI futures, Speculative Design, and Design Fiction methodologies.

RESOURCES FOR HCI PRACTITIONERS AND DESIGNERS

1. UNESCO Futures Literacy
   - What You Get: Access to educational materials, research papers, and tools for developing Futures Thinking and literacy skills.
   - Link: https://en.unesco.org/futuresliteracy
2. OECD Future of Education and Skills 2030
   - What You Get: Insights into future trends shaping education and skills development, along with policy recommendations.
   - Link: https://www.oecd.org/education/2030-project/
3. Houston University's Future Studies Program
   - Description: Dive into the Future Studies Program at Houston University, offering cutting-edge research, courses, and forums for exploring emerging technologies and their implications on society.
   - Link: https://www.houstonforesight.org/
4. Tsinghua University Future Lab
   - What You Get: Cutting-edge research, publications, and events on future studies and human-computer interaction (HCI).
   - Link: https://thfl.tsinghua.edu.cn/en/index.htm
5. MIT Media Lab
   - What You Get: Access to interdisciplinary research, projects, and events exploring the intersection of technology, media, and design.
   - Link: https://www.media.mit.edu/
6. Turku University Futures Studies
   - Link: https://www.utu.fi/en/study-at-utu/masters-degree-programme-in-futures-studies

     Description: Access Turku University's interdisciplinary resources on futures studies, including academic programs, research publications, and networking events fostering collaboration in envisioning alternative futures.
7. Google Design Sprint Kit
   - What You Get: A comprehensive guide and toolkit for running design sprints to solve complex problems efficiently.
   - Link: https://designsprintkit.withgoogle.com/
8. GitHub Open Source Design
   - What You Get: Templates, tools, and resources for open-source design collaboration and community building.
   - Link: https://github.com/opensourcedesign

9. World Economic Forum's Future of Work Resources
   - Description: Engage with the World Economic Forum's curated content on the future of work, featuring insights on automation, digitalization, and skills development for the workforce of tomorrow.
   - Link: https://www.weforum.org/agenda/2020/01/the-future-of-work-look-like-2030/
10. Dubai Futures Foundation
    - What You Get: Workshops, courses, and resources on Futures Thinking, innovation, and emerging technologies.
    - Link: https://www.dubaifuture.ae/
11. Global Futures Database (GFD) – People Share
    - Description: Connect with a global network of futures thinkers and practitioners on the Global Futures Database (GFD), facilitating knowledge sharing, collaboration, and co-creation of innovative solutions to complex societal challenges.
    - Link: https://gdf.network/network/

4 Future of Entertainment Technology Design

Nandhini Giri and Erik Stolterman

INTRODUCTION

This chapter discusses course pedagogy and the authors' experiences of experimenting with futures studies in facilitating discussions on the future of entertainment technology design with students specializing in computer graphics technology design that includes animation, game design, data visualization and human-centered experience design. The classroom research activities are intended to apply futuristic design methodologies in existing pedagogical curriculum to challenge students to think about their own entertainment professional practices, grounded in a future where digital entertainment is intertwined with social, cultural, environmental, and political factors. We provide examples of our experimentation with fictional narratives and futuristic design methodologies in classroom and research settings.

Emerging entertainment technologies are redefining the notion of immersive storytelling, enabling media consumers to take up more active roles in designing personalized experiences that are meaningful to their lived experiences. Artificial intelligence enabled media platforms, real-time applications of computer graphics entertainment and emerging conversations on the metaverse are frequently discussed pedagogical topics in educational environments. This shifting landscape in the entertainment technology space necessitates a future-oriented design pedagogy in course curriculum, classroom learning experiences and research mentoring. The first example provides an analysis of student's imaginary narratives of digital entertainment futures in the metaverse. The objective of this study is to understand student's perception of the future of metaverse experiences, emerging entertainment technology design and transactional services. The idea was to allow students to critically imagine potential futures of non-fungible experiences and design human interactions with emerging entertainment technologies. The activity involved undergraduate students traveling to the year 2050 and account for their futuristic visions. Students documented their immersive experiences in a future-oriented metaverse that connects their current lives to a time in the future. The fictional narratives were analyzed to build generic design patterns that highlight interesting questions on what the future of non-fungible experiences could be and how it can shape our lives in a distant time. Our inspiration for this activity comes from the concept of future cones (Hancock & Bezold, 1994; Voros, 2001), Burdick's process of designing the future "from the inside" (Burdick, 2019) and Baerten's article on napkin futures (Baerten, 2019). The second example provides an account of an ongoing graduate-level class project about applying speculative design in game development projects. Students in a graduate course explore the future of college sports in the metaverse and develop

DOI: 10.1201/9781032693606-6

interactive prototypes that embody the on-campus sports culture, while projecting the sports spirit into futuristic digital platforms. Finally, the article concludes with our experiences of mentoring graduate students in future-oriented research topics. Novel entertainment technologies promise new research opportunities but lack evidence of prolonged success and impact among its stakeholders. We discuss the challenges and opportunities in exploring futuristic research topics and the lessons learnt through these classroom activities and research mentoring experiences. We share what we consider necessary mentoring tools to equip students with the necessary skills to critically think about the future and conduct impactful research to shape the whole ecosystem of entertainment technology futures.

BACKGROUND LITERATURE AND MOTIVATION FOR EXPERIMENTATION

The work on futures studies by Voros (2001), lays the foundations for our pedagogical experiments with futurology in understanding the future of entertainment technology design. The opening statement in Voros' primer explains that the master concept of the futures field is in the existence of many alternative futures instead of one simple future. Futures work is built upon the three "laws" of future, in that the future is not predetermined, the future is not predictable and the future outcomes can be influenced by the choices we make in the present. This foundational concept motivates our classroom experiments with futurology. The recent advances in the field of entertainment technology and the shift in media consumption and interaction behaviors of users – due to the impact of the pandemic have necessitated these futurology discussions in classroom with students trained in computer animation, games, and interaction design.

The four classes of alternative futures adapted from Henchey (1978) namely the possible, plausible, probable, and preferred futures correspond to the questions of what "might" happen, what "could" happen, what "is likely to" happen and what "we want to" happen. These alternative ideations of the future are dependent on the current knowledge, trends, possibilities, and value judgments that result from our collective informational, cognitive, and emotional sources. We decided to explore the pedagogical implications of futures studies by applying these concepts in classroom assignments.

Design education lays a lot of emphasis on the design process, and the learning outcomes often focus on improving student's skills in mastering design tools and applying design principles. Futures studies area has the potential to add the layer of critical thinking in student work that allows them to think about the implications of their design work, the long-term impact of their design products and how these critical ideas impact their own design process. Burdick's article (Burdick, 2019) of designing futures from the inside provides an account of imagining new practices through the simultaneous creation of story worlds, plots, and narrative elements with an emphasis on relations as opposed to things. The author suggests that this method of exploring human-concerned futures from the inside – "design fiction creators as participants-observers" explores the interconnectedness of the futures by the internal worlds of our daily lives. The storytelling element in Burdick's work and the idea of building fictional narratives through imagined lived experiences is yet another piece that we wanted to experiment with in our classroom activities.

In Sterling's work (2013) there arises the question of what the atomic element of a design fiction is – the least thing that one can do which constitutes a design fiction. This pondering over the atomic element of design fiction led to the exploration of Baerten's work (2019) on napkin futures. Baerten refers to archaeological practices of digging up fragments of past worlds and cultures, to actively reflect on the bigger picture by uniting the fragments in another timeframe. The suggestion to apply the same practice in the engagement of life in a futuristic context motivated our experiments. Napkin futures or micro narratives sketched on paper napkins appear to be stories from the future. The main take-away from this article for our own classroom exercises is the author's perspective on moving away from dichotomy-based thinking to a way of thinking, acting, and knowing based on diversity.

PEDAGOGICAL EXPERIMENTS AND CONTRIBUTION TO FUTURES STUDIES

The following sections provide our documentation of sample pedagogical experiments conducted in coursework over a period of three semesters with undergraduate, graduate, and doctoral research students specializing in computer graphics and interactive entertainment technologies. Each activity applied the philosophy of Voros' future cone approach in identifying the four future classes, foresight, and future scenario building. There was an emphasis on the design process with futuristic design products tied to our daily lives and envisioned from within with the fictional narrative creator being a simultaneous observer of their own creation. Finally, a continuous mentoring attempt was made to help students think beyond a logical exchange of narratives and be able to see the interconnected emergent worlds from these fragmented narratives. The motivation, activities, insights from each case study contributes to the scholarship of applying futures studies in the context of designing entertainment technologies of the future.

A summary of these pedagogical experiments and their contributions are listed below:

Experimentation # 01: Speculating the Future (s) of Entertainment Technologies

Objective: Apply futures thinking concept in freshman course to speculate futuristic entertainment technologies and experiences
Activities: Design fiction, Storyboarding
Insights: Qualitative analysis of student work helped identify futuristic themes that question humanity's ideas of entertainment and interaction with these technologies

Experimentation # 02: Sports Futures in the Metaverse

Objective: Develop futuristic prototypes that embody college sports culture
Activities: Trend analysis, foresight, plausible futures, and what if scenarios

Insights: Hands-on approach helped students prototype futuristic solutions and critically analyze plausible futuristic scenarios

Experimentation # 03: Futurology in Student Research Work and Mentoring

Objective: Critically analyze the implications of student design practices
Activities: Student mentoring, project advising
Insights: Ambiguity of evaluation methods for futuristic solutions, siloed narratives confined to student's field/discipline of study, lack of student's personal lived experiences in narratives

SPECULATING THE FUTURE(S) OF ENTERTAINMENT TECHNOLOGIES

Our first example documents the application of speculative design and futures thinking in a first-year classroom environment. The course introduces students specializing in computer graphics technologies to the fundamentals of graphics programming, visual design, motion graphics and human-centered design. The course is structured as modules with each module focusing on the different foundational concepts of graphics and entertainment technologies. During the final weeks of the semester, students were given a lecture on futures thinking in class. The lecture content explained how design thinking process diverges – starting from the current situation but converges into a feasible user-centric product. Futures thinking on the other hand, diverges and leads to the possible, plausible, probable, and preferred futures as explained by Voros' future cone concept. Speculative design approach was introduced to students from Dunne and Raby's book (2013) on speculative everything, highlighting the fact that speculative design challenges design, to imagine alternative futures and pose problems for critical ideation. This speculative approach shifts design practices from applications to the implications end. The quantum parallelogram (StudioPSK) and the ice-cream cloud project (Tucknott, 2009) were provided as examples to these approaches. The introductory content was followed by some warm-up exercises that include the "brain cap" activity for students to imagine future technologies that provide unlimited resources in regard to knowledge, emotions, experiences/memories, skills and messaging. Another follow-up activity was the "Thing from the future" card game (Candy & Watson, 2015) that prompts students to ideate on different future/thing/theme cards. The overall involvement of students with the lecture content was positive and interactive.

This class lecture and activity was followed by an online assignment in which students were encouraged to speculate future-oriented experiences in the metaverse and human interactions with emerging entertainment technologies. The assignment's objective was to understand students' perceptions of how emerging technologies are shaping the future of entertainment, especially in the context of the metaverse. Students were encouraged to apply the futures cone to speculate possible, plausible, probable, and preferred futures and generate futuristic narratives situated in the Summer of 2050.

Here are the assignment instructions provided to students:

1. Fixate on one future (probable, plausible, possible, preferable). Imagine an entertainment product or service in this alternative future. Situate it in a relevant social, political, environmental, or cultural context. Storyboard human interactions with this product or service.
2. Write a short narrative about your recent metaverse experience from Summer 2050 (max 500 words)

The student submissions were analyzed to develop themes and identify patterns in the fictional threads that explain how they perceive immersive experiences in a future-oriented metaverse that connects their current lives to a time in the future. The key take-away from this exercise is an understanding of what students in the computer graphics technology field of study perceive the future definition of entertainment activities could be and how interwoven it looks with our everyday lives. The following subsections provide details of the various themes analyzed in the student work highlighting some recurring examples.

FUTURES: DIGITAL ENTERTAINMENT EXPERIENCES

A major theme observed in students' speculations of future-oriented entertainment experiences is the personalization and diversification of digital entertainment experiences. The narratives explored futures in which humans have complete control over the content they consume and how it is consumed. The fictional narratives referred to how advanced technological intervention makes our everyday lives more engaging. The ubiquitous nature of these entertainment platforms is intertwined with human's daily activities making entertainment more educational, functional, and supportive of enhancing the quality of everyday human emotions. The pandemic's influence is also evident in the speculations made by students about the future trends that shape these digital entertainment experiences. The following recurring examples were embedded in the fictional narratives and storyboards.

- Travel bots for cultural literacy: Narratives explored technological interventions in virtual tourism with the assistance of immersive technologies and travel bots as tourism guides to cultural destinations. "Once you are on a camel, you can feel the breeze, smell the sand and see the scorching sun high up in the sky." Students were optimistic about the photo-realistic graphical renderings in delivering naturally lived experiences, but also attributed these virtual travels as an alternative to counter extreme climate changes and withering cultural heritage sites.
- Spiritual retreats and mental rejuvenation: Many examples highlighted the issues of prolonged media effects and the need for mental rejuvenation and spiritual retreats within virtual worlds. One of the speculated solutions was a digital detox from the multi-worlds using an immersive technology shower. Figure 4.1 shows a visual sketch of this concept.

FIGURE 4.1 Visual sketching of futuristic experiences with entertainment technologies.

- Reliving childhood memories: This narrative was interesting, especially looking at one's own life as a source of entertainment by revisiting the past and enacting the role to re-experience events through stored memories. "I went back to my 8-year birthday party where the theme was 'puppies' and got to experience eating the best cake of my life again … It freaks me out a little bit because it feels weird knowing everyone in your memory is experiencing it for the first time, but this is your second, third, or even tenth time reliving it." Figure 4.2 shows a visual sketch of the concept.

FUTURES: SHARED DIGITAL EXPERIENCES

The narratives focused primarily on personal interactions with entertainment technologies. However, they also provided a scope for shared digital experiences. References were made to existing technological affordances and how it can support shared experiences in futuristic digital platforms. Some of the multimodal interaction examples included holographic performance balls that project musical performances in social settings, drones with immersive projectors for shared experiences, smart

FIGURE 4.2 Visual sketching of futuristic experiences with entertainment technologies.

contact lenses and touch screen windows. Students also speculated about interactions with celebrities and influencers in these futures "You could tell a story of yourself or a story of how the celebrity influences you. If the celebrity found your story interesting enough, then you would have a chance to meet them in the metaverse." Crypto farms and Non-Fungible Tokens (NFT) museums were other speculated venues for shared digital experiences. Some of the narratives also described transactions in the metaverse, used for virtual travel and souvenir collection – "In one day, I was able to see the Eiffel Tower, the Leaning Tower of Pisa, the Colosseum, Notre Dame, and Big Ben with my family. I also did a lot of shopping with my friends too! We were able to get a ton of cool virtual outfits by selling some NFTs."

FUTURES: METAVERSE AND ENTERTAINMENT TECHNOLOGIES COLLECTIVE DESIGN

This subsection takes the examples and emerging themes from the previous section to investigate some of the design considerations to shape the future. This includes human factors for prioritization of digital experiences and the ramifications of these futuristic technologies. A major portion of the fictional narratives explored futuristic entertainment experiences through the lenses of existing technologies like drones, holographic displays, sensors, and wearables. The speculations were situated in altered futures in which digital entertainment experiences are influenced by social, economic, technical, and environmental factors. Fictional narratives showed the influence of multiple stakeholders and interconnected systems in the design of entertainment technologies, products, and services. A need for self-care and questions of privacy were voiced by students as they critically analyzed the implications of their design solutions for the future of entertainment. The activity and the student work reduce to the question of whether the future of digital entertainment is going to be an introvert's world?

SPORTS FUTURES IN THE METAVERSE

The second example provides an account of applying speculative design in a graduate course that explores the future of college sports in the metaverse and developing interactive prototypes that embody the on-campus sports culture, while projecting the sports spirit into futuristic digital platforms. This semester long activity was divided into multiple components to engage students to think about the various design aspects of the final prototype. Further, these speculative designs were prototyped in an online metaverse platform to demonstrate a proof of concept of how college sports viewing experiences can be enhanced in emerging graphics technology platforms.

The ideation of these futuristic prototypes began with an exploration of traditional entertainment practices. Students were provided with the prompt to explore traditional practices that form a source of entertainment within their own cultural context. Students presented their exploration on this topic responding to prompts on: (1) What is this traditional practice about? (2) Why it is entertaining? (3) What are the design elements that make this experience entertaining? (4) Do you identify and theorize the practice with the whole experience? This activity yielded discussions on a diverse source of entertainment experiences that inspire people from various cultural traditions. Classroom presentations and discussions revolved around the similarities in entertaining patterns and the different approaches employed in the practice.

The next exercise was related to experiencing realities. Students were tasked with the problem of framing a new reality and make their peers experience this reality and later talk about how they design this reality. The topic was abstract and triggered a lot of questions from students on what is expected of this class assignment. The results were quite interesting as each student came up with their own interpretation of what they see as an experience in a reality defined by the individual. Presentations included multimodal approaches to experiencing these framed realities. A follow-up to this activity was to identify entertainment modalities. This time students were to (1) identify an entertainment system in the physical – digital continuum, (2) discuss the input and output modalities of this system and user interaction modes, (3) explain how the system's affordances improve or decrease the user's interactive experiences, and (4) make suggestions to improve the interaction experience with the chosen system.

The final activity in this series was to explore emerging technologies and to speculate future oriented tools and interfaces. The task was to (1) identify emerging trends in each individual student's field of study related to technology tools and interfaces, (2) use these trends to speculate the future of technical tools and interfaces in this field over a span of 5 years, and (3) discuss the influence of these technological trends on human experience and interactions in everyday lives. Students coming from various backgrounds ranging from virtual reality technologies, user experience, animation and social gaming presented work across multiple fields and technologies. Each of these activities and classroom presentation were spread across the semester and helped structure student's ideas and approach to prototyping the sports future within the metaverse.

The sports future in the metaverse prototype also followed a set of design activities and the process influenced the way the final prototypes were delivered by the students. Students identified trends in the University sports culture that included historical facts, student rituals, statistics and blog posts related to the sports experiences among students. These trends provided directions to predict plausible futures of the University sports culture especially for the basketball and football teams. This trend analysis also helped students consolidate the major factors that influence these plausible futures. The next step was to develop four scenarios by picking four major factors from the trend analysis. This scenario development was followed by students providing design guidelines for developing futuristic stadium sports experiences in the metaverse.

This part of the assignment was graded based on the (1) validity of the data sources used in the trend analysis of college sports and culture, (2) coherence of trend prediction and scenario development and characterization, and (3) practical application of the prescribed design guidelines. This individual section of the metaverse sports futures assignment introduced students to a structured format to approach speculative design and predict future trends. This exercise also helped them gather the necessary preliminary data for the teamwork section of the project. Students were divided into two teams (basketball and football) and further collected data about the assigned sports team, on-campus student experiences and online activities related to sports events. Student teams identified pain-points from the data analysis to discover opportunities that could enhance the sports experience in the metaverse. The discovery activity led to speculations of the future to build a metaverse presence for college sports that is a solution space for the current issues faced on-campus. The prior activities described in the initial part of this section were a guideline to prototype interaction scenarios for in-world (metaverse) sports experiences. The prototypes were further tested out in a commercial metaverse content development platform through a sponsored company project.

This semester long project resulted in students speculating future-oriented prototypes for college sports experiences in the metaverse. These speculated concepts were also tested out in a real-time metaverse content generation platform. This hands-on approach has helped students develop the necessary skills in critical ideation, prototyping futures by building upon the four future-oriented ideation exercises and to present the working prototypes to a professional company (that sponsored the projects) for feedback and validation of their concepts.

FUTURES STUDIES IN STUDENT RESEARCH WORK AND MENTORING

A general observation from the above two examples in the above sections is that there is an increasing interest among students to experiment with emerging technologies. There is greater interest and engagement with hands-on activities and explorative assignments that challenge student's mindsets. This section discusses the challenges of mentoring and advising students in these classroom activities., We notice professional titles like "metaverse expert," "web3.0 consultant," "technology futurists," and more. The pedagogical experiments documented in this article have not necessarily trained students in becoming these experts. However, the methods and discussions have provided them with necessary tools to critically analyze their own design process and evaluate the impact of their work in building the future of entertainment technology.

The main challenge faced in these experiments is the lack of an evaluation rubric that can provide consistent feedback on student's work. Futures thinking and critical reasoning were major learning outcomes of each of the classroom activities. However, we faced the challenge of evaluating design solutions – does a plausible future scenario of a decentralized metaverse service prove to be more impactful than a preferred one? Are the foresights developed for futuristic services in the music industry reliable enough to pursue? The corporate world seems to apply these tools to stay ahead of their competitors in terms of designing future products from existing data and trends. In an educational setting, there is greater freedom to test out new ideas in a sandbox. But the ambiguity of the evaluation methods urges us to seek more tangible well-tested frameworks to provide better feedback to students of futures studies. Another challenge we faced in these classroom activities is the interdisciplinary nature of futures studies. Students were trained in programming, human-centered design, and visual language. Ideation sessions and presentations revealed narratives that were siloed within the student's field of study. It was also difficult to identify simple narratives in the speculations. Students were excited about their knowledge of science fiction movies and video games. Their fictional narratives were a lot influenced by these popular storylines and culture. It took some conscious effort in identifying personal narratives that came from inside, through each individual's lived experiences.

The reference section of this chapter includes a list of scholarly articles that were part of the graduate course reading list and class discussions. The article on games as speculative design (Coulton, Burnett, & Gradinar, 2016) was of particular interest, given the student's educational background. The article's discussion on the plurality of perception of the past, present, and future helped generate open conversations on the alternative futures. Framing games as a medium of speculation and iteration provided step-by-step guidelines for students to generate ideas and iteratively connect them to co-create a complex ecosystem from simple individual narratives. One plausible future of wearable entertainment technologies and a preferred future scenario of body ingestible trackers sparked conversations of a collective future co-created by the two discussed scenarios.

CONCLUSION

Overall, our experiments with futures studies in the classroom have been positively engaging and have been a process of discovery and learning. Articles on futures studies, speculative design, design fiction, probes and world building mandala have opened new doors for futuristic exploration. These experimentations can be applied in classrooms and curriculum related to entertainment graphics technology design. The insights gained through these experiments contribute to futuristic design themes that question student's perception of entertainment technologies and human interactions in the future. They provide an opportunity to critically analyze the implications of student design practices. Insights also highlight a need for better evaluation methods of these futuristic solutions and ways to improve interdisciplinary work and personal narratives in classroom discussions of futures studies. We plan to explore further literature in this space and use games and interactive visualization software as a speculative platform to test out futuristic narratives. We plan to develop objectively and subjectively characterized futuristic probes and evaluation frameworks for intervention in the ideation, prototyping, development, and testing phases of entertainment product design in future classroom activities.

REFERENCES

Baerten, N. (2019, June). Retrieved from https://jfsdigital.org/wp-content/uploads/2019/06/12-Baerten-Napkin-Futures.pdf

Burdick, A. (2019). Designing futures from the inside. *Journal of Futures Studies, 23*(3), 75–92.

Candy, S., & Watson, J. (2015). *The thing from the future.* The APF methods anthology. London: Association of Professional Futurists.

Coulton, P., Burnett, D., & Gradinar, A. I. (2016). Games as speculative design: allowing players to consider alternate presents and plausible features. In Lloyd, P. and Bohemia, E. (Eds.), Future Focused Thinking - DRS International Conference 2016, 27–30 June, Brighton, United Kingdom. https://doi.org/10.21606/ drs.2016.15

Dunne, A., & Raby, F. (2013). *Speculative everything: Design, fiction, and social dreaming.* Cambridge, MA: MIT Press.

Hancock, T., & Bezold, C. (1994). Possible futures, preferable futures. *The Healthcare Forum Journal, 37*(2), 23–29.

Henchey, N. (1978). Making sense of futures Studies. *Alternatives, 7,* 24–29.

Sanders, E. B. N., & Stappers, P. J. (2014). Probes, toolkits, and prototypes: Three approaches to making in codesigning. *CoDesign, 10*(1), 5–14.

Sterling, B. (2013, November 26). Design fiction: Napkin futures. Wired. Retrieved from http://www.wired.com/2013/11/design-fiction-napkin-futures/

Studio PSK. Quantum Parallelograph. Retrieved from https://www.studiopsk.com/quantum parallelograph.html

Tucknott, T. (2009). Edible Cloud Project. Retrieved from https://www.trendhunter.com/trends/the-cloud-project

Voros, J. (2001). A primer on futures studies, foresight and the use of scenarios. *Prospect: The Foresight Bulletin, 6*(1), 1–8

Yantaç, A. E., Vatansever, A., Omaç, T., Kayı, İ, & Kuşcu, K. (2021, June). WorldBuilding and Mandala as a Tool for Co-Speculating on the Healthcare Domain in 2050. In *Proceedings of the 24th International Academic Mindtrek Conference* (pp. 81–90).

5 Long-Term Thinking in Digital Product-Service System Design

Yanru Lyu and Bowen Zhang

INTRODUCTION

The transition toward digital Product-Service Systems (PSSs) is a significant challenge for manufacturing industries. In addition, the advent of Industry 4.0 and the impact of the pandemic have highlighted the need for new skills and strategies to address the changing social, economic, and professional landscape. PSS, which combines products and services, provides a promising solution to address environmental and economic issues, but its uptake across industries is limited (Dyah et al., 2023). Due to digitalization and the commoditization of hardware, the need for innovative business models that integrate offerings and form business ecosystems has appeared (Humbeck et al., 2022). The development of PSS in business ecosystems requires organizational capabilities and a framework for future development (Humbeck et al., 2022). Transforming the PSS strategy toward digitalization involves conceptual design, prototyping, and analyzing the impact on the process of customized digital service design. Future strategies, business models, and organizational capabilities should be focused during the transition toward digital PSS. SMEs need more resources, expertise, and technological infrastructure. Therefore, methods and tools that support SMEs in adopting digital PSS are essential.

Long-term thinking, also referred to as foresight strategy, long-term versioning, or future-oriented thinking, is the practice of intentionally considering what may happen in the future (von Weizsäcker, 2019). This is vital for digital PSS toward effective transition and sustainable development. While there is some research on tools based on a long-term perspective in PSS (Lelah et al., 2014), a gap exists in understanding how to incorporate long-term thinking into implementation for digital transformation.

The work presented in this study follows a dual strategy that combines theory and practice. The literature review led to the emergence of a theoretical framework, while field studies were carried out with the industry case.

LITERATURE REVIEW

The move to digital PSS means taking advantage of new technologies and business models to offer value-added services and solutions to customers. Long-term visioning can assist companies in identifying opportunities for innovation and growth during the digital transition. This section covers a literature review of the central concepts involved.

DOI: 10.1201/9781032693606-7

Moving to Digital PSS

PSS can be defined as an integrated solution that is composed of products and services to meet customer needs (Mont, 2002), concentrating on the life cycle of products and services and promoting sustainable development through the circular economy (Baines et al., 2007; da Costa Fernandes et al., 2020). Tukker (2004) divided PSS into product-oriented, use-oriented, and result-oriented directions and eight types based on the degree of servitization. Product-oriented services are still primarily geared toward product sales, but some extra services have been added. Use-oriented services are not geared toward selling products but are sometimes shared by several users. Finally, result-oriented services deliver a functional performance without any pre-determined product involved. During the design process, PSS is described as a challenging task that has to work seamlessly with traditional product design (Rondini et al., 2016); additionally, several domains need to integrate from the later phases of the life cycle into the early stages of the design process and raise the demand for methods and strategies that support collaboration and cross-disciplinary integration in design (Yan et al., 2022). Moreover, customer requirements analysis is also a critical part of the design, beginning to ensure that the designed solution aligns with the needs and preferences of the target customers (Lyu et al., 2022).

Digital PSS integrates various digital technologies (e.g., the internet of things, cloud computing, blockchain, 3D technology, virtual reality, augmented reality, artificial intelligence (AI), and machine learning) to provide products and services for fulfilling customer needs in a fast-changing and the digitalized world (Richter et al., 2019). In other words, integrating digital technologies would provide customers with innovative and efficient solutions that cater to their evolving needs in a rapidly changing digital landscape. There are various digital transformation types for PSS based on the digital technology profile of the company, the direct or indirect transformation of physical to digital ones, and ecosystem expansion issues (Kim, 2021). As digital transformation becomes a global mega-trend topic for many enterprises, novel approaches are becoming increasingly required (Lerch & Gotsch, 2015).

Based on design thinking, the stage to construct digital PSS involves defining the problem, concept development, prototyping, testing, and renewal. In the classic process of PSS design, testing the market and consumer demand at the beginning is essential (Rondini et al., 2016), which can be the engine to drive the concept development. Prototypes are conducive to experiencing and discovering potential problems (Yan et al., 2022). Testing and renewal will gradually make the final plan near the goal.

Long-term Perspective and Approaches

A long-term perspective is of great necessity in various fields, including PSS design. *Foresight* is a systematic and multidisciplinary environmental data analysis and interpretation process focused on building future scenarios that support organizations' decision-making (da Rosa & Janissek-Muniz, 2023). As the specific subset of foresight, technology foresight emphasizes predicting and shaping future technological developments and their impacts on society, economy, and industry, which is becoming essential in dealing with uncertainties from a long-term perspective.

Besides, analysis is also indispensable in PSS design as it is conducive to identifying early indicators of potential future trends and disruptions that may affect the industry.

There are different methods that could be employed to forecast future trends. The Foresight Diamond is a framework that positions methods based on their primary type of knowledge sources (i.e., creativity, expertise, interaction, or evidence) (Popper, 2008). In addition, he also explored the potential contribution of 33 methods for decision-making and transformation. Literature Review and Scanning can be performed during the pre-foresight stage to gather more information and insights about current trends and developments. A series of methods, including Brainstorming, Expert Panels, and Scenarios, would contribute to more generation. Surveys and road mapping are more efficient methods for action inspiration that could provide valuable guidance. Furthermore, an Interview is an effective way for renewal.

Digital Technologies

Artificial Intelligence

Different methods could be employed to forecast future trends. Generative AI, such as text-to-image and image-to-image models, has gained widespread use, mainly for its efficiency and quality improvements (Lyu et al., 2023). The traditional design paradigm would be changed. In marketing, AI has been used to improve advertising campaigns, collect user data, and create advertisements that are perceived differently from those created by humans (Jia, 2023). In fashion, customized solutions involve various AI techniques like neural networks and genetic algorithms (Chan et al., 2022). Moreover, it has also driven innovation in business models and enhanced organizational capabilities (Astafeva et al., 2020).

Future AI systems will be able to communicate with human users and each other and efficiently exchange knowledge and wisdom with abilities of cooperation, collaboration, and even co-creating something new and valuable and possess meta-learning capacities (Cichocki & Kuleshov, 2021). AI can be used to solve complicated issues in the future. For example, it can help data adjust their energy usage automatically, also be used to predict the demand for energy usage in advance by combining renewable and non-renewable energy, and can be utilized to analyze Big Data for security breaches (Gill et al., 2022).

Blockchain

The blockchain is a linked set of records maintained in a decentralized environment, where the records are publicly available but cryptographically secured; in addition, once information is recorded, it is infeasible to modify it (Priyadarshini, 2019). It is a radical rethinking of how we pay for things and verify who owns them and who has the right to buy or sell them (Kostin, 2018). Currently, a large number of blockchain applications are found in the financial sector, and an increasing number of applications in different fields are emerging (An et al., 2023). Timestamp and immutability are critical for applying the blockchain in the cultural sector (Clemente & Bifulco, 2023). In the industrial internet, blockchain technology can promote the transformation and upgrading of the industrial system and contribute to the evolution of the industrial system toward modernization and intelligence (Li et al., 2022). In the future, the blockchain will be applied to more fields to support the development of the field or enhance the system in the field.

Big Data

Big data deals with the dynamic, extensive, and disparate volume of data created by different parties, including people, tools, and machines (Kostin, 2018). It allows businesses to compile information from various sources, create knowledge, and enhance predictions and business innovation (Mao et al., 2022). Due to the use and exploitation of data, it is possible to make intelligent decisions when proposing strategies that are better targeted and adapted to a specific target audience and objectives specific to the organization (Rivera & González, 2022). In the context of supply chain credit assessment for small- and medium-sized enterprises (SMEs), financial service providers (FSPs) utilize big data analytics (BDA) to identify the quality and potential risks of borrowers and provide tailored information, aiming to help SMEs obtain financing (Song et al., 2021). In the evaluation decision-making of PSS alternatives, stakeholder comments can be analyzed using big data to construct an index system and determine the value of PSS alternatives (Li & Mao, 2020). Here, it should be noted that the evolution of BDA is driven by advancements in data resolution, coverage, and velocity, as well as the variety of datasets. Machine learning methods, such as deep neural networks, are becoming more potent in analyzing big data (Lutfiani & Meria, 2022).

There is a mutual promotion between technologies. In other words, developing one technology can also trigger acceleration in developing another technology, such as AI assisting in extensive data analysis.

THE SUSTAINABLE DIGITAL PSS DESIGN FRAMEWORK

Connecting digital PSS design and foresight methods to sustainability involves considering each design stage and matching it with foresight methods. The aim is to ensure that any decision-making process during digital PSS design is focused on short-term needs and issues and long-term sustainable innovation and foresight. The sustainable digital PSS design workflow based on foresight methods is presented in Figure 5.1. Under the influence of market, profit, and context, technology can be derived to realize the digital transformation of PSS, while foresight methods are constantly assisting design toward sustainable future innovation. Below, we elaborate on each framework stage, detailing the foresight methods used and their integration into the design process.

Define the Problem

In this initial stage, it is essential to clarify the current situation and future challenges the SME faces by identifying global issues, pinpointing core values, and detecting opportunities linked to the question (Lelah et al., 2014). In general, problems are mainly defined in narrow terms of short-term innovation and user needs in the

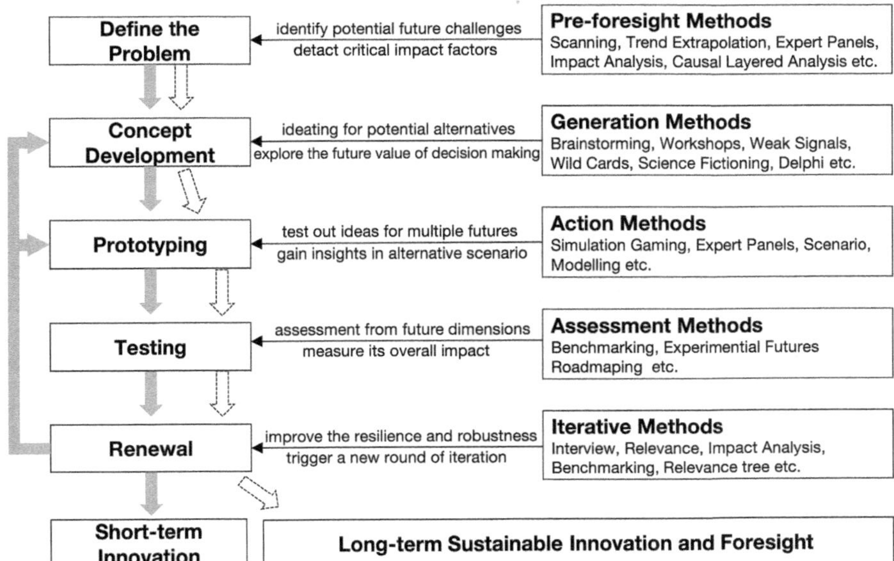

FIGURE 5.1 The sustainable digital PSS design framework.

background of technology development. Environmental scanning involves surveying and interpreting relevant data to identify early indicators of potentially essential developments (Choo, 2001). It detects weak signals and emerging trends that may impact the organization's strategic direction. In addition, expert panels can bring together specialists from diverse fields to provide valuable insights and feedback. This approach fosters a holistic understanding of the problem and its potential future implications (Linstone & Turoff, 1975).

Concept Development

By envisioning multiple possible futures, one problem may be related to a potential alternative. Brainstorming, as a creative group technique, aims to generate various ideas to solve problems (Osborn, 2012). Within this framework, it explores potential alternatives for addressing the identified problem. Besides, workshops focusing on free-thinking activities can elicit perspectives on possible futures, broaden viewpoints, and foster creative solutions (Jungk & Müllert, 1987). Wild Cards, which consider low-probability, high-impact events, could significantly alter future scenarios and encourage thinking about unexpected futures and their potential impact on the digital PSS (Petersen, 1999).

Prototyping

Rapid prototyping simulates real experiences and tests ideas for multiple futures. Scenario planning creates multiple, preferred future scenarios to test prototypes against (Schwartz, 1997). In this framework, exploratory scenarios start from current trends and develop forward to describe several plausible futures. Simulation gaming can also explore future contexts and test prototypes within those frameworks (Duke, 2014).

Testing

The testing phase incorporates foresight methods to ensure comprehensive economic, social, and environmental assessment. As a structured communication approach, the Delphi method can be used to develop and refine assessment criteria (Linstone & Turoff, 1975). Trend impact analysis combines statistical extrapolation with qualitative judgments about the effects of possible future events (Gordon, 1994). These methods assist in evaluating the potential impact of the digital PSS prototype across different dimensions.

Renewal

Based on the feedback from testing, the renewal stage triggers a new round of brainstorming and decision-making. As a planning method, backcasting begins with defining desirable futures and then works backward to identify policies and programs connecting that specified future to the present (Robinson, 1990). This method helps refine the digital PSS prototype based on long-term sustainability goals. Roadmapping, on the other hand, is a flexible technique that supports strategic and long-range planning by matching short-term and long-term goals with specific technology solutions (Phaal et al., 2004). It is used to plan the implementation and future development of the digital PSS.

FIELD STUDIES

Applied to Fashion Companies

Two fashion companies upgrading their digital PSS were taken to illustrate the design framework. Company A is an SME that provides long-term B2B clothing design services. This project can be characterized by the transformation of the offer design, the business model, the internal skills, and the organization. Unlike that, company B has developed a B2C product that can be more efficient with a PSS offer, which provides consumers with unique jewelry styles. Table 5.1 underlines some features of the digital PSS design process. The reason for choosing these two companies as samples is because they both belong to SMEs, which provide fashion design services and products, with differences in scale and customer. We collaborated with them and used foresight methods to design the digital PSS for transformation.

Company A currently offers standard clothing designs and faces pressure due to a lack of creative designers, the need to keep up with fashion trends, and the necessity to communicate with producers effectively. By scanning, the company's strengths include a positive market reputation, stable customer resources, and abundant design capabilities. External environment trend analysis reveals generative AI technology value in design and a growing demand for sustainable materials and

TABLE 5.1
Underlines Some Features of the Strategic Decision Situations

	Company A	Company B
Market	B2B	B2C
Employees	30–40 people	Less than 10 people
Customers	Clothing producer	Jewelry consumers
Product offer	Clothing samples	Jewelry
Service offer	Design new clothing styles	Customize the preferred style of jewelry
Situation	• Pressure comes from a lack of creative designers • Keep up with fashion trends • Communicate with producers	• Cost much communication time • Be difficult to extend scale • Keep consumer loyalty
Define	Internal Environment Analysis: • Market reputation • Stable customer resources • Rich design resources External Environment Analysis: • Emerging trends of AI design • Declaration on sustainable materials • Growth for original creativity Potential questions in the futures: • Potential risks of reduced demand due to AI design • Environmental degradation and labor exploitation caused by the fast fashion	Internal Environment Analysis: • Brand reputation • Fans resources • Organizational flexibility External Environment Analysis: • Emerging trends of AI design • Digital consumption trends • Personalized demand growth Potential questions in the futures: • Scarcity of precious metals • Consumption in the metaverse
Development	• Trend prediction based on big data • Collaboration through AI channel • Technical services at any time	• Material and craft upgrade • Communication by generative AI • Asset library building
Prototyping	Build an AI fashion design system tool to achieve automated design	Purchase API services for quickly generating 2D and 3D renderings
Testing	Invite designers and producers to test	• Invite customers to test
Renewal	System and business model update • Starting from clothing, and then expanding to other categories • More recommendations for sustainable materials • The gradual transformation from software and hardware to software services	System and business model update • Automatically filtering through aesthetic calculations • Recycling and reuse mechanism • Delayed consumption by obtaining NFT

original creativity. Through brainstorming sessions with experts and meticulous study of weak signals, potential future scenarios have been identified, encompassing a possible decline in demand due to the adoption of AI-driven design and the imperatives of addressing ecological concerns and labor exploitation within the fast fashion industry. During the development stage, the system design needs to avoid problems effectively. Therefore, the wild card methods that involve brainstorming with experts and analyzing weak signals to identify potential future scenarios and develop strategies to navigate discussion focus on trend prediction based on big data, collaboration through AI channels, and instant technical services. Then, the first vision of AI-driven fashion system prototypes was built for experience testing from economic, social, and environmental dimensions. Based on the potential future issues, sustainable materials, and step-by-step transformation were highlighted for the update.

As for Company B's B2C market segment, it caters to jewelry consumers by offering customized jewelry designs. The current situation was scanned to show high costs and communication time, challenges in scaling the business, and maintaining consumer loyalty. In the internal environment analysis, fundamental factors involve brand reputation, fan resources, and organizational flexibility. On the other hand, external environment analysis highlights emerging trends in AI design, digital consumption patterns, and the growth of personalized demand. Due to the small scale of this company, it is not easy to invite experts, considering that it is more suitable to engage stakeholders in some fictional imagination. In the future, human beings will be forced to explore new materials because of the shortage of nonrenewable metals. At the same time, consumers can first digitize asset storage to meet their ownership needs while delaying the acquisition of physical goods. This is a very environmentally friendly idea. Outsourcing and purchasing API services are ideal for small team operations without a stable technical team. By inviting customers to test, the assessment can combine the user's experience and sustainable perspective. After a few rounds of discussion, it triggered a new round of brainstorming and decision-making, such as adding an automatic filter function, recycling and reuse mechanism, and digital twin value.

Advantages and Challenges

Based on the results and the company's feedback, there are many advantages and potential values as follows:

Comprehensive Analysis: The framework allows for a thorough analysis of internal and external factors that could impact the design and implementation of digital PSS, leading to a more informed decision-making process.

Strategic Planning: By identifying strengths, weaknesses, opportunities, and threats, companies can develop strategic plans to leverage their strengths and address weaknesses in designing digital PSS.

Alternative Innovation: The framework encourages companies to explore new technologies, trends, and potential scenarios, promoting innovation in the design of digital PSS.

Below are some potential challenges that arise from observation and reflection.

Risk Mitigation: Companies can proactively address risks and uncertainties in designing and implementing digital PSS by considering potential challenges and wild cards. However, these resources are often limited for SEMs.

Integration Complexity: Integrating digital technologies, services, and products into a coherent PSS may be complex and challenging, requiring careful exploring alternative integration paths and comparing their advantages and disadvantages.

Execution Challenges: These challenges stem from various factors, including ambiguity in the rule of design and uncertainty in the future. To improve the robustness of the plan, more stakeholders would be involved in the co-creation for desirable futures.

CONCLUSIONS

Foresight methods can be followed to guide a company, especially an SME, to clarify its vision and make the right decisions to support the transition to digital PSS and align with a sustainable strategy adapted to its specific context. The sustainable digital PSS design workflow guides the construction of new PSS in the digital age. The choice will reflect on sustainability and value co-creation that best responds to customers' needs and environmental and social goals.

This chapter explored how to engage in the construction of digital PSS in line with a long-term perspective was explored. A series of foresight methods were examined to integrate into the design process. Combining the workflow, a case study conducted in parallel on two SMEs supported the propositions. The research findings indicate that a long-term perspective and foresight approach can link to sustainable innovations such as digital transformation in a company. Even in SMEs, these light and convenient tools can help organizations quickly iterate design and assist in decision-making. Although they may encounter limitations such as funding and resources, internal flexibility can effectively address potential external challenges.

REFERENCES

An, M., Fan, Q., Yu, H., & Zhao, H. (2023). Blockchain technology research and application: a systematic literature review and future trends. *arXiv preprint arXiv:2306.14802.*

Astafeva, O., Pecherskaya, E., Tarasova, T., & Korobejnikova, E. (2020). Digital transformation in the management of contemporary organizations. *Digital Age: Chances, Challenges and Future* (pp. 382–389). Springer International Publishing.

Baines, T. S., Lightfoot, H. W., Evans, S., Neely, A., Greenough, R., Peppard, J., Roy, R., Shehab, E., Braganza, A., & Tiwari, A. (2007). State-of-the-art in product-service systems. *Proceedings of the Institution of Mechanical Engineers, Part B: Journal of Engineering Manufacture, 221*(10), 1543–1552.

Chan, L., Hogaboam, L., & Cao, R. (2022). Artificial Intelligence in Fashion. In L. Chan, L. Hogaboam, & R. Cao (Eds.), *Applied Artificial Intelligence in Business: Concepts and Cases* (pp. 325–334). Springer International Publishing. https://doi.org/10.1007/978-3-031-05740-3_21

Cichocki, A., & Kuleshov, A. P. (2021). Future trends for human-AI collaboration: A comprehensive taxonomy of AI/AGI using multiple intelligences and learning styles. *Computational Intelligence and Neuroscience, 2021*, 1–21.

Clemente, L., & Bifulco, F. (2023). Time Stamp and Immutability as Key Factors for the Application of Blockchain in the Cultural Sector. In *Handbook of Research on Blockchain Technology and the Digitalization of the Supply Chain* (pp. 310–326). IGI Global.

Choo, C. W. (2001). Environmental scanning as information seeking and organizational learning. [version électronique] *Information Research, 7* (1), 29

da Costa Fernandes, S., Pigosso, D. C., McAloone, T. C., & Rozenfeld, H. (2020). Towards product-service system oriented to circular economy: A systematic review of value proposition design approaches. *Journal of Cleaner Production, 257*, 120507.

da Rosa, L. M., & Janissek-Muniz, R. (2023). The impact of foresight on digital transformation strategies. *Revue internationale d'intelligence économique, 14*(1), 63–83.

Dyah, W. K., Himawan, R., & Ardianwiliandri, R. (2023). Strategy framework development based on business commercial integrating co-creation process to guide PSS transition. *JEMIS (Journal of Engineering & Management in Industrial System), 11*(1), 60–83.

Duke, R. D. (2014). *Gaming: the future's language*. wbv Media GmbH & Company KG.

Gordon, T. (1994). Trend impact analysis. In J. C. Glenn & T. J. Gordon (Eds.), *Futures research methodology* (pp. 1–21).

Gill, S., Xu, M., Ottaviani, C., Patros, P., Bahsoon, R., Shaghaghi, A., & Uhlig, S. (2022). AI for next generation computing: Emerging trends and future directions. *Internet of Things, 19*, 100514.

Humbeck, P., Rosenfelder, J., & Bauernhansl, T. (2022). Organizational Capabilities for the Development of PSS in Business Ecosystems. 2022 Portland International Conference on Management of Engineering and Technology (PICMET), Portland, OR, USA. IEEE.

Jia, Y. (2023). Application and development of artificial intelligence based on computer science. *Applied and Computational Engineering, 6*, 874–878. https://doi.org/10.54254/2755-2721/6/20230947

Kim, Y. S. (2021). Digital transformation types for product-service systems. *Proceedings of the Design Society, 1*, 1283–1292.

Kostin, K. B. (2018). Foresight of the global digital trends. *Strategic Management-International Journal of Strategic Management and Decision Support Systems in Strategic Management, 23*(1).

Lelah, A., Boucher, X., Moreau, V., & Zwolinski, P. (2014). Scenarios as a tool for transition towards sustainable PSS. *Procedia CIRP, 16*, 122–127.

Lerch, C., & Gotsch, M. (2015). Digitalized product-service systems in manufacturing firms: A case study analysis. *Research-Technology Management, 58*(5), 45–52.

Li, L., & Mao, C. (2020). Big data supported PSS evaluation decision in service-oriented manufacturing. *IEEE Access, 8*, 154663–154670.

Li, Z., Zhang, B., Li, J., Guo, J., & Nie, F. (2022). Analysis of Industrial Internet Application Based on Blockchain. 2022 International Conference on Information Technology, Communication Ecosystem and Management (ITCEM),

Lutfiani, N., & Meria, L. (2022). Utilization of big data in educational technology research. *International Transactions on Education Technology, 1*(1), 73–83.

Lyu, F., Feng, Z., Li, F., & Su, J. (2022). Customer requirements analysis in PSS design.

Lyu, Y., Shi, M., Zhang, Y., & Lin, R. (2023). From image to imagination: Exploring the impact of generative AI on cultural translation in jewelry design. *Sustainability, 16*(1), 65.

Linstone, H. A., & Turoff, M. (Eds.). (1975). *The delphi method* (pp. 3–12). Reading, Addison-Wesley.

Mao, Z., Li, H., Huang, Z., Tian, Y., Zhao, X., & Zhang, H. (2022). Research on the value and application of power big data in external value-added services. 2022 International Conference on Cloud Computing, Big Data Applications and Software Engineering (CBASE),

Mont, O. K. (2002). Clarifying the concept of product–service system. *Journal of Cleaner Production, 10*(3), 237–245.

Osborn, A. (2012). *Applied imagination-principles and procedures of creative writing*. Read Books Ltd.

Popper, R. (2008). Foresight methodology. *The handbook of technology foresight*, 44–88.

Petersen, J. L. (1999). *Out of the blue: How to anticipate big future surprises* (2nd ed.). Madison Books.

Priyadarshini, I. (2019). Introduction to blockchain technology. *Cyber security in parallel and distributed computing: concepts, techniques, applications and case studies*, 91–107.

Phaal, R., Farrukh, C. J., & Probert, D. R. (2004). Technology roadmapping—a planning framework for evolution and revolution. *Technological forecasting and social change*, 71, 5–26.

Richter, A., Glaser, P., Kölmel, B., Waidelich, L., & Bulander, R. (2019). A review of product-service system design methodologies. *ICETE (1)*, 121–132.

Rivera, M. R., & González, K. V. (2022). Advantages and Benefits of Big Data in Business Communication. In *Marketing and Smart Technologies: Proceedings of ICMarkTech 2021, Volume 1* (pp. 279–292). Springer.

Rondini, A., Pezzotta, G., Pirola, F., Rossi, M., & Pina, P. (2016). How to design and evaluate early PSS concepts: The product service concept tree. *Procedia CIRP, 50*, 366–371.

Robinson, J. B. (1990). Futures under glass: a recipe for people who hate to predict. *Futures, 22*(8), 820–842.

Song, H., Li, M., & Yu, K. (2021). Big data analytics in digital platforms: How do financial service providers customise supply chain finance? *International Journal of Operations & Production Management, 41*(4), 410–435.

Schwartz, P. (1997). *Art of the long view: planning for the future in an uncertain world*. John Wiley & Sons.

Tukker, A. (2004). Eight types of product–service system: Eight ways to sustainability? Experiences from SusProNet. *Business Strategy and the Environment, 13*(4), 246–260.

von Weizsäcker, E. U. (2019). Science and long-term thinking—The club of Rome, a club of long-term thinkers. *Europhysics News, 50*(2), 29–31.

Yan, Z., Larsson, T., & Larsson, A. (2022). PSS value transformation: From mass-manufactured vehicles to provision of mass-customized services—A case study of designing and prototyping customized digital services for SAIC motor in China. *Proceedings of the Design Society, 2*, 1179–1188.

Section III

Competence and Practice

6 Design Futures Methodologies

Qing Xia

INTRODUCTION

In recent times, the temporal and spatial range of design has continuously widened. Temporally, it extends from tomorrow to the anthropocene; spatially, it evolves from "objects" to "systems that include objects," and even to the formulation of macro policies and exploration of cultural forms and philosophical paradoxes. As content expands, design method research inevitably interacts with other future-oriented studies, adopting methodologies from other fields to augment its own toolkit. On the other hand, with technological advancement, design can leverage big data and AI to rapidly deliver products or services tailored to user "tastes," shifting the design subject from designers to immediate user needs. This has led to relatively short-sighted design outcomes with convergent features. As a response, many researchers are turning to other fields for supplemental thinking patterns and methodologies. These factors have collectively contributed to the richness and increased quantity of methods and tools in designing futures. These methods and tools are primarily used for idea generation and visualization, communication and collaboration, efficiency enhancement, exploration and experimentation, evaluation and testing, historical and cultural reflection, future orientation.

Since Herbert Simon defined the purpose of design as "problem-solving" (Simon, 1996), mainstream design methods have followed a functionalist path. Due to the pursuit of feasibility and commercialization, current mainstream design methodologies mainly focus on conceiving products and services for the near future, such as double diamonds, design thinking, etc. In this short-term future zone, authors like Bella Martin, Bruce Hanington (Martin and Hanington, 2012), Marc Stickdorn (Stickdorn, Hormess and Lawrence, 2018), and Robert A. Curedale (Curedale, 2013) have summarized design tools from different perspectives like design stages, categories, and methodologies. However, methods for conceiving mid- to long-term futures and their systematic analysis are relatively sparse. In recent years, methodologies for mid- to long-term futures, such as speculative design (Dunne and Raby, 2013) and design fiction, have emerged. These methodologies often borrow methods and tools from other fields like strategic foresight, technological foresight, and future studies, and have proven effective in practice. Yet, the design methods for this time frame lack comprehensive organization and need further summarization regarding their applicability. Currently, designers mainly imitate the tool usage of renowned design institutions or universities, which lacks specificity and flexibility. Addressing these issues, this study organizes and summarizes the methods and tools already applied in designing futures, and further discovers potential tools from a broader range of

DOI: 10.1201/9781032693606-9

fields. Building on this, the study treats designing futures as a general creative practice process, overall arranging future design methods and those from other fields for combined practical application.

Based on the objectives, the research content of this chapter is divided into three main parts:

Extensive collection of existing tools. The study selects tools not limited to those that inherently reflect the future or temporality, but that can be used in future design practice.

Verification of tool effectiveness. Tools found are categorized into two groups: those already widely used or proven effective in future design work are directly included in the study's toolkit; potential tools, unverified, are added after validation through a mini-Delphi method.

Distribution of methods and tools according to the design process stage. Tools collected are arranged in the order of use within the general process of creative practice, facilitating joint application by designers.

The specific research path of this study follows an "identification-evaluation-classification" process as seen in Figure 6.1.

During the evaluation phase, for methods not yet verified in practice, this study organized a mini-Delphi method consisting of 12 experts to assess these methods as shown in Table 6.1. Through two rounds of email communication questionnaires,

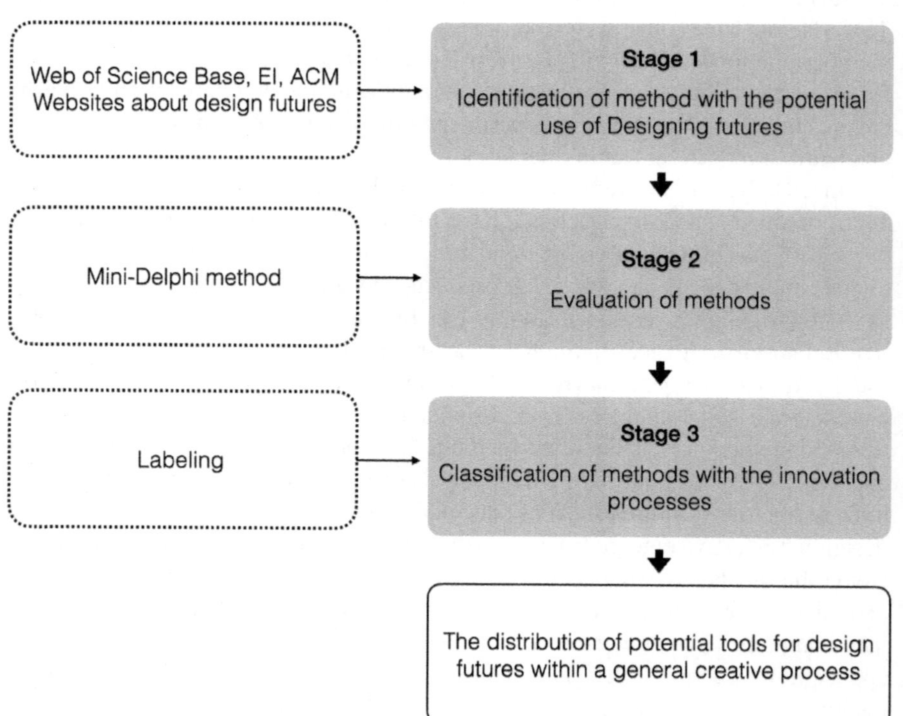

FIGURE 6.1 Research process.

TABLE 6.1
Background Information of Experts Participating in the Mini-Delphi Method

Characteristic Group	Expert Distribution (Number of Experts)
Research Field	Interaction Design: 4, Service Design: 3, Business Design: 2, Product Design: 2, Environmental Design :1
Profession	Design Education and Research: 6, Design Practice: 4, Design Consulting: 2
Years of Experience	5–10 years: 6, 10–20 years: 6

experts were asked to give a comprehensive score on the potential application of these tools in the context of designing futures. The experts used a four-level assessment scale to determine the extent to which a particular method supports design futures practice, where: 0 – indicates unsuitability, 1 – indicates low applicability, 2 – indicates average applicability, and 3 – indicates high applicability.

The collected results were analyzed. Subsequently, the rating scale for the evaluation of methods with potential use in design futures, together with the results from round 1, was again sent to the same experts. In the next round of the study, the experts completed the same rating scale, while having the opportunity to familiarize themselves with the aggregated results from the first round of the study. They were able to compare their own positions with the opinion of the group, and after analyzing the arguments could change their mind. The questionnaires received from the second round of the study have been re-examined. Eventually, a list of potential tools for designing futures has been obtained.

In the classification stage, the study organized an analytical team composed of six design method researchers as shown in Table 6.2. The team first determined a general creative process that could encompass both the design process and other future exploration processes. Subsequently, they tagged the methods in the toolkit according to which stage of the creative process they are best suited for.

Through this "identification-evaluation-classification" process, the study derived a list of methods that can be used for futures design practice. This list aids designers in finding and jointly using tools appropriate for the stage of design they are in, advancing the design process, adding dimensions to thinking, and ensuring the quality of design outputs.

TABLE 6.2
Personnel Information of the Classification Stage Analysis Team

Characteristic Group	Research Personnel Distribution (Number of Researchers)
Position	Associate Professors: 2, Lecturers: 3, PhD Candidate: 1
Research Direction	Design Futures: 2, Design Methods: 3, Interaction Design: 1
Years of Engagement in Design Methods Research	10–20 years: 3, 5–10 years: 2, 1–5 years: 1

DESIGN-ORIENTED DIFFERENT TIME ZONES IN THE FUTURE

Design has always been about the future. Whether we're creating products, services, or policies, there's a temporal span from identifying a problem to implementing a solution, where design introduces new entities into a previously non-existent world. Thus, all design is future-oriented (Candy, 2010). For the task of designing futures, the time span presets macro background factors like society, technology, economics, environment, and policy, and fundamentally affects the creators' mindset. Theories like Construal Level Theory (Förster, Friedman, and Liberman, 2004) and the concept of Hypothetical distance (Wakslak and others, 2006) reveal how the perception of near versus distant futures and the temporal proximity affect expression. This study investigates design research's main content and methodologies across different future time spans.

To study design patterns across various time spans more clearly, the study adopts future research and design management approaches to segment future time into zones based on a natural generational scale (Bühring and Liedtka, 2018), dividing into four periods: within 5 years, 5–20 years, 20–50 years, and beyond 40 years as seen in Figure 6.2.

The near future, targeted by existing product and service designs, mainly focuses on the immediate needs of users and market conditions within the next five years. Utilizing current or soon-to-be-matured technologies, the aim is to rapidly produce and capture the market for swift monetization. Given the tight timelines for realization, this design category is significantly constrained by practical factors, making it suitable for confirmatory design driven by perfection and closely tied to actual production activities. Design within this time span is largely inspired and limited by design thinking and business logic, starting from problems to empathize with people's current needs.

The mid-term future is where design institutions and tech organizations present "future visions" and research entities release "future technologies." The commonality in these designs is the use of scenario-based presentations to depict convenient, novel, and improved lifestyles enabled by smart products. These designs encourage concrete associations from the present to the future rather than detailing specific functions and benefits. Designs in this timeframe are influenced by constant human needs, long-term institutional planning, and technological trends. Current methodologies in this

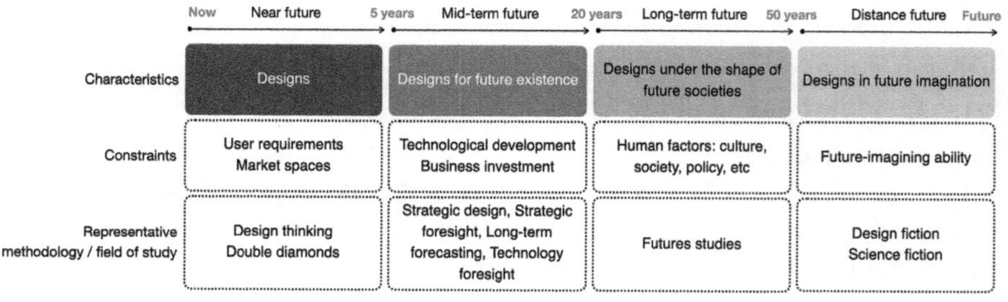

FIGURE 6.2 Design in different future zones.

space still rely on design thinking, which is actionable but limited in envisioning beyond current possibilities. Supplements like Peter Scupelli's Dexign futures and Terry Irwin's Transition design explore new methods for this zone, offering new mental models for the mid-term future. This phase allows for exploring stable trends while providing room for necessary change, encompassing both feasibility and possibility. It's a focus of interdisciplinary research between design and futures studies (Day and Schoemaker, 2006), encompassing methods like environmental scanning for identifying long-term drivers of change and organizations like Pantopicon that explore the uncertainties of design for 5–15 years into the future (Pantopicon, 2022). Design for this timeframe benefits from current and future-oriented methodologies, broadening the scope for conceptualizing future smart products.

The long-term future focuses on conceiving smart products within specific societal forms, representing a microcosm of future society. Design for this horizon, spanning 20–60 years, is less about the design itself and more about its role in narrative; these designs serve as "diegetic prototypes," a term coined by Bruce Sterling, provoking thought about alternative solutions to real-world problems. Such designs, which speculate on societal constructs, prioritize the portrayal of a world's beliefs and operational principles over detailed depictions. Designers for this future need to think broadly and be sensitive to trends, as the design reflects macro factors like cultural traditions and societal forms. Speculative design is the prevalent approach, with futures studies methodologies being increasingly integrated into the design process.

Designs aimed at distance future often emerge in science fiction novels, films, paintings, and other artistic works, as well as a few cases within design civilization. Compared to long-term future designs, these are broader in thought, imagining entities never before seen in our civilization, striving to inspire and feel plausible. The main constraints are the limits of experience and imagination. These designs typically serve as accessories to a larger background story, setting the tone of a future world rather than narrating a story on their own. So far, models, methods, and tools for envisioning such products are rare, with some explorations in sci-fi, such as World Building (McDowell, 2019).

From the analysis of the four future time periods, we can discern patterns that show: (1) Design influences shift from individual to societal, and even to civilization; (2) Design expressions move from functional details to conceptual backgrounds, with forms ranging from realistic to abstract to "pseudo-real"; (3) Design concepts evolve from conforming to demands to reflecting, questioning, and even critiquing the present; (4) Design methodologies transition from de-futurization to futurization, that is, from realizing a single future to seeking diversity of futures; (5) Design methods and tools expand from within the discipline of design to multiple (hybrid) domains.

DESIGN METHODS FOR DIFFERENT TIME ZONES

Due to the incremental nature of design across different future time zones, this study individually examines tools for designing futures in these zones. For the near future, tools are reviewed mainly within the design field; for the mid-term future, reviews encompass design, strategic foresight, long-term forecasting, and technology foresight; for the long-term future, apart from some conceptual and cutting-edge design research, reviews are mainly in future studies; and for visionary futures, tools are examined in imaginative fields like design fiction and sci-fi. It's noted that most tools aren't specifically devised for a particular future zone, with their categorization based on usage frequency, and many tools can apply across multiple periods. This classification is limited by statistical representation and doesn't imply exclusive applicability to any stage (as seen in Table 6.3).

During the organization of potential tools for designing futures, experts predominantly excluded some quantitative research tools from long-term forecasting and technological foresight. Post-mini Delphi, experts were interviewed via email about their reasons for low ratings. The tools were not chosen as potential future design tools due to: (1) high specialization and learning costs for designers; (2) easily accessible results or data from these tools, which designers could directly reference; (3) insufficient effectiveness for design purposes.

THE DISTRIBUTION OF POTENTIAL TOOLS FOR DESIGN FUTURES WITHIN A GENERAL CREATIVE PROCESS

Based on a wide collection of potential tools for designing futures, this study aims to further analyze which stage of the actual creative process these tools should be used in. This analysis helps designers conveniently combine various tools to advance the design process. Given the diverse origins of these tools, not only from design but also from strategic foresight, long-term forecasting, decision-making, technological foresight, futures studies, and sci-fi, the study broadens the conception of the design process to a more inclusive model of the creative process. This creative process, defined as a series of thoughts and actions leading to original and appropriate outputs (Lubart, 2001), is described at two levels: the macro, characterized by various stages of the process, and the micro, explaining the mechanisms behind it. To facilitate interdisciplinary comparisons, this chapter primarily analyzes the macro process.

Through summarizing various creative processes, we observe overlaps between disciplines despite their unique aspects (Botella and Lubart, 2015). Initially, processes are merged due to similarity, mostly around "identifying necessary problems to solve" (Osborn, 1963) also known as problem selection (Busse and Mansfield, 1980), sensitivity to the problem (Guilford, 1956), or problem definition. In some disciplines, a pre-identification discovery stage exists, like the "empathy" stage in design thinking or Torrance's "perceiving gaps" stage (Torrance, 1962). Additionally, Mumford and others suggest scrutinizing the problem during discovery, rejecting unrealistic, incorrect, or incomplete issues (Mumford, Reiter-Palmon, and Redmond, 1994).

Following is the preparation stage, initially described by Wallas in his early macro process model (Wallas, 1926). As Carson explains, this stage involves defining or understanding the problem and gathering information to solve it (Carson, 1999; Treffinger, 1995). Some scholars also identify a subsequent incubation stage. According to Wallas and others, incubation is a period where thoughts and associations occur subconsciously, between conscious and unconscious levels (Osborn, 1963). Considering the lack of external actions during incubation, this study merges

TABLE 6.3
Potential Tools for Design Futures

Use of Design for Near Future			
$100 test	Graffiti walls	Participant observation	The love letter and the breakup letter
5-second usability test	Guided tour	Personal inventories	Usability testing
A/B testing	Heuristic evaluation	Personas	User interview
AEIOU	How Now Wow matrix	Ring of experience	User journey map
Crowdsourcing	Intercept interview	Service blueprint	User observation
Diary study	KANO Analysis	Shadowing	Value proposition map
Empathy map	Lean canvas	Simulation exercise	Wizard of Oz
Fly-on-the-wall	Paper prototype	Stakeholders	WWWWWH
Use of Design for Mid-Term Future			
5 whys	Morphological analysis	Force field analysis	Citizen panel
Concept mapping	Branch analysis	FMEA	Technological scanning
Business model canvas	Horizon scanning	STEEPVL	MANOA
Business origami	Environmental scanning	Structural analysis	Requirements analysis
Concept testing	STEEP	Cross-impact analysis	Analogue forecasting
Culture probes	Trend impact analysis and reviews	Strategic position and action evaluation	Estimating multi perspective
Experience prototype	11 macro sources of disruption	Assessment of the impact on society	Trees references
Fishbone	Roadmapping	Sensitivity analysis	Megatrend analysis
Flexible prototype	Technology mapping	Cost-benefit analysis	Trend and impact analysis
PEST	Technology scouting	Institutional analysis	State of the future index
Provocation	Tech mining	Factor analysis	Randomized controlled
Role-playing	Technology readiness levels	Analysis of the action	Numerical process
SWOT analysis	Key technologies	Correspondence analysis	Ensemble forecasting
Territory map	Life cycle analyses	Risk analysis	Scenario planning
Value opportunity analysis	Analysis of long-term	Sustainability analysis	The red team simulation
Scenario Description Swimlanes	S-curve analysis	Classification trees	Moral algebra
Mapping the IoT toolkit	Trend extrapolation	Analytic hierarchy process	Emotional census
Living futures: scenario kit	Benchmarking	Cluster analysis	Regulatory impact analysis
What the Block?	Time series analysis	Probability trees	Linear numerical modeling
Symbiosis tool	Retrospective analysis	Correlations	The bad event
Relevance trees	ANKOT	Ranking lists	Weighted matrix
Use of Design for Long-Term Future			
Diegetic prototype	Causal layered analysis (CLA)	Future workshops	Structural analysis

(Continued)

TABLE 6.3 *(Continued)*
Potential Tools for Design Futures

Future scouting	Charrette	Futurecasting	The futures polygon
Futures-design-process model	CIPHER	Futures landscape	Thought experiment outside limited perspective
Possession Tool	Contradictory evidence	Futures polygon	Three horizons
Speculative design tools	Delphi	Futures triangle	Transcend method
Video prototype	Emerging issues analysis	Influence diagram	Verge
Vision toolkits	Extreme options	Jury verdicts	Visioning
4-quadrant mapping	FTI forecasting	Multi-level perspective	Visual images
Archetypes	Future cone	Nuts and bolts	Wild cards
Axes of uncertainty	Future signs	One day in the future	–
Backcasting	future wheel	Shared history	–
	Use of Design for Distance Future		
Design fiction	World Building	Thought experiment	–
The thing from the future	Science fiction	–	–
	No Specific Application Zone, Used with Other Tools		
2×2 matrix	Design jam	Lotus blossom diagram	Sketch Prototype
6 thinking hats	Design mash-up	Mind mapping	Speed dating
Affinity diagramming	Design sprint	Mood board	Storyboard
Brainstorming	Dot voting	Point-of-view statement	World café
Card sorting	Fast idea generator	Questionnaire	Assumptions vs Knowledge
Conversation starter	Focus group	SCAMPER	Expert panel
Creative workshop	HMW How might we?	Scenario	Storytelling
Design brief	Image board	Secondary research	Sketch Prototype

* The gray background show methods are rarely applied in design futures.

the pre-creation preparation and incubation stages for a more inclusive process that accommodates various types of practice.

The third stage is the core of creative activity—the creation phase. Here, creators define constraints related to the problem and transform these into outcomes (Busse and Mansfield, 1980). Some scholars propose a subsequent narrowing process, the selection stage, where creators choose from multiple options (Bruford, 2015). Others suggest a creative development stage after creation and selection, for reimagining and reorganizing ideas (Mace and Ward, 2002). This study integrates the processes of creation, filtering, and recreation into the creation stage.

After creation comes the deployment of the idea, known as the application phase (Gruber, 1989). Treffinger explains this stage as leading to action through planning (Treffinger, 1995). During this stage, the solutions found can be implemented and realized.

Finally, creators test and validate the realized output, which is the evaluation stage (Runco, 1997; Wallas, 1926). This phase involves assessing the chosen and applied ideas, deciding whether to abandon, delay, store, destroy, execute, or produce them (Carson, 1999). Additionally, some experts mention a post-application communication and reflection stage to discuss the pros and cons of the application and reflect on its potential consequences (Botella et al., 2013; Cropley and Cropley, 2012). This study integrates post-application testing, validation, evaluation, communication, and reflection into the evaluation stage.

Based on the analysis above, this study establishes a "problem identification-preparation-creation-application-evaluation" process for creative activities. Building on this, the study utilized an analysis team consisting of six researchers who labeled the existing tools to determine the stage applicable for each tool. This approach allows for a comparative presentation of commonly used heuristic design tools and tools from futures studies as seen in Figure 6.3. In the figure, tools that are widely used in the design process are positioned in the upper half, while tools that are not yet widely used in the design field are placed in the lower half. Additionally, tools that can be applied to multiple stages are positioned in the boxes at both ends of the image.

By mapping potential tools for design futures onto a general creative process, we can more clearly identify the design stages each tool is suited for. This analysis enables designers to more conveniently combining various tools to advance the design process. Additionally, our analysis revealed that: (1) There is already close methodological collaboration between tools from design and other future-related fields, with potential for further mutual learning; (2) The creation and application stages are crucial for the visualization strengths of design tools, while preparation and evaluation stages benefit from in-depth, broad analytical capabilities of other futures-related research areas; their combination helps to address each other's limitations; (3) The availability of potential methods varies across stages in terms of quantity, categories., and applicability. Some stages offer a wider range of methods, while others have limited options. Future research could further identify stages with fewer available tools, assess unmet needs, and evaluate the necessity of developing new tools.

DISCUSSION

This study, from a usage perspective, extensively collected potential tools for designing futures across different future time horizons. These tools were then categorized and mapped onto a general creative process model using a labeling approach. During the research process, we identified several potential avenues for further analysis.

1. Different research perspectives can lead to varying results in tool selection and classification. The core objective of this study is to assist design practitioners in the design futures practice by enabling them to select and use tools more conveniently and effectively. Therefore, when selecting and classifying tools, the primary considerations are their usage frequency and practical application across stages, rather than the inherent features of the tools or the original intentions of their designers. Analyzing existing tools from the latter perspectives might yield a different set of tools.

Methods that have proven effective in design futures

Methods that can be used in any process

Design sprint
Design jam
Futures-design-process model
Speculative design tools
Mapping the IoT toolkit
Living futures: scenario kit
Creative workshop
World cafe
Future scouting
What the Block?

Problem Identification

Secondary research
Graffiti walls
Stakeholders
Conversation starter
User Observation
Shadowing
Fly-on-the-wall
Guided tour
Simulation exercise
User interview
Focus group
Intercept interview
Questionnaire
Personal inventory
Diary study
Culture probe
Crowdsourcing
Card sorting

Preparation

Personas
Empathy map
User journey map
Value proposition map
HMW How might we?
Point-of-view statement
Territory map
Mood board
Image board
Design brief

Creation

Brainstorming
Lotus blossom diagram
Mind map
Design mash-up
6 thinking hats
Fast idea generator
SCAMPER
Fishbone
Provocation
Affinity diagram
2x2 matrix
Business model canvas
Lean canvas
PEST
SWOT
5W1H
PAPCM
How Now Wow
KANO model
Value opportunity analysis
$100 test
Weighted matrix
Dot voting
Possession tool
Speed dating
The thing from the future

Application

Tomorrow headline
Scenario
Storyboard
Paper prototype
Flexible prototype
Technical documentation
Video prototype
Role-playing
Wizard of Oz
Experience prototype
Blueprint
Business origami

Evaluation

Concept testing
Usability testing
Heuristic evaluation
A/B testing
5-second usability test
Net promoter score
Semantic differential scale
Feedback capture grid
5E model
The love letter & the breakup letter
System usability scale
Testing design using the competition
Snap tests

Problem Identification

Shared history
Futures triangle
Futures polygon
Futures landscape
Multi-level perspective
Emerging issues analysis
future wheel
Future signs
Future cone
Relevance trees
Morphological analysis
Branch analysis
Horizon scanning
Verge
Technology mapping
Technology scouting
Tech mining
Weak signals
Morphological analysis
Technological scanning
Retrospective analysis
Requirements analysis

Preparation

Causal layered analysi
Structural analysis
4-quadrant mapping
11 macro sources of disruption
Environmental scanning
SWOT
PEST
STEEP
Three horizons
The futures polygon
Axes of uncertainty
Trend impact analysis and reviews
Assumptions vs Knowledge
Analysis of long-term
Life cycle analyses
S-curve analysis
Benchmarking
Time series analysis
ANKOT
Trend extrapolation
Force field analysis
FMEA
STEEPVL
Structural analysis
Citizen panel
Social networks analysis
Megatrend analysis
Technology readiness levels
Technology barometer
Trees references
MANOA
State of the future index
Influence diagram

Creation

Nuts and bolts
Scenario
Futurecasting
Storytelling
Thought experiment
Archetypes
Visioning
Expert panel
Technology road mapping
Analogue forecasting
Estimating multi perspective
Visions of the future
Wild cards
Future workshops
Charrette
Jury verdicts
Contradictory evidence
Thought experiment
Extreme options
Randomized controlled
Numerical process
Ensemble forecasting
The red team simulation

Application

One day in the future
Visual images
Backcasting
Transcend method
Roadmapping
Key technologies
The bad event

Evaluation

Factor analysis
Analysis of the action
Institutional analysis
Correspondence analysis
Risk analysis
Cluster analysis
Sustainability analysis
Classification trees
Probability trees
Correlations
Ranking lists
Assessment of the impact on society
Cross-impact analysis
Strategic position and action evaluation
Life cycle assessment
Cost-benefit analysis
Sensitivity analysis
Trend and impact analysis
Moral algebra
Emotional census
Regulatory impact analysis
Linear numerical modeling

Methods that can be used in any process

Delphi
CIPHER

Methods that are rarely applied in design futures

FIGURE 6.3 Distribution of potential design futures tools in the general creative process.

2. The model for the design futures process exhibits potential diversity. To accommodate methods from various fields more broadly and expand the range of tools available to design futures practitioners, this study generalized the design process. This approach may, to some extent, compromise the distinctive features of the design futures work in the process. Employing different model strategies in the research could more specifically yield design futures processes tailored to particular goals.
3. Further research is needed on how to integrate potential tools for design futures. The current study is limited to a relatively macro-level categorization of tools for each stage. However, understanding how to combine and apply these tools in practice is a critical issue that must be thoroughly explored before using them in design futures work.

CONCLUSIONS

As the scope of design expands in terms of time and space, and as technology introduces issues of short-sightedness and convergence, the methodologies for design futures have become increasingly diverse, with a growing number of methods. Currently, there is a relatively rich body of research on methods for short-term futures, but fewer studies address methods for envisioning medium- to long-term futures beyond this time frame. To address this gap, this study organizes and summarizes the methods and tools already used in design futures practice and identifies potential tools from broader fields that could be adapted. Based on this, the study treats design futures as a general creative practice process and outlines potential methods for handling futures, facilitating their combined application in practice.

This study used an "identification-evaluation-classification" process. In identification phase, existing tools are extensively collected.. During the evaluation phase, a mini-Delphi method is used, with 12 experts assessing the tools collected in the previous phase in two rounds to determine which methods should be retained. In the grouping phase, potential methods for design futures are mapped onto a general creative process model to aid practitioners in selecting and applying them.

Through this process, the study has generated a list of tools for design futures. This list enables designers to more easily combine various tools and advance the design process. Building on this study, further research could identify stages with insufficient methods, explore the development of new methods to meet unmet needs, and investigate how to integrate potential tools for design futures effectively.

REFERENCES

Botella, M., Glaveanu, V. P., Zenasni, F., Storme, M., Myszkowski, N., Wolff, M., & others. (2013). How artists create: Creative process and multivariate factors. *Learning and Individual Differences*, 26, 161–170.

Botella, M., & Lubart, T. (2015). Creative processes: Art, design and science. In G. E. Corazza & S. Agnoli (Eds.), *Multidisciplinary contributions to the science of creative thinking* (pp. 53–65). Springer.

Bruford, W. (2015). Making it work: Creative music performance and the Western kit drummer. University of Surrey. https://billbruford.com/wp-content/uploads/2021/11/Bruford-Thesis.-Version-of-Record.pdf

Busse, T. V., & Mansfield, R. S. (1980). Theories of the creative process: A review and a perspective. *The Journal of Creative Behavior*, 14(2), 91–103, 132.
Bühring, J., & Liedtka, J. (2018). Embracing systematic futures thinking at the intersection of strategic planning, foresight, and design. *Journal of Innovation Management*, 6(3), 134–152.
Candy, S. (2010). The futures of everyday life: Politics and the design of experiential scenarios. https://doi.org/10.13140/RG.2.1.1840.0248
Carson, D. K. (1999). Counseling. In M. A. Runco & S. R. Pritzker (Eds.), *Encyclopedia of creativity* (Vol. 1, pp. 395–402). Academic Press.
Cropley, D. H., & Cropley, A. J. (2012). A psychological taxonomy of organizational innovation: Resolving the paradoxes. *The Creativity Research Journal*, 24, 29–40.
Curedale, R. A. (2013). *Design methods 2: 200 more ways to apply design thinking*. Design Community College.
Curedale, R. A. (2013). *Service design: 250 essential methods*. Design Community College.
Day, G. S., & Schoemaker, P. J. (2006). *Peripheral vision: Detecting the weak signals that will make or break your company*. Harvard Business Press.
Dunne, A., & Raby, F. (2013). *Speculative everything: Design, fiction, and social dreaming*. The MIT Press.
Förster, J., Friedman, R. S., & Liberman, N. (2004). Temporal construal effects on abstract and concrete thinking: Consequences for insight and creative cognition. *Journal of Personality and Social Psychology*, 87(2), 177–189.
Gruber, H. E. (1989). The evolving systems approach to creative work. In D. B. Wallace & H. E. Gruber (Eds.), *Creative people at work: Twelve cognitive case studies* (pp. 3–24). Oxford University Press.
Guilford, J. P. (1956). Structure of intellect. *Psychological Bulletin*, 53, 267–293.
Lubart, T. I. (2001). Models of the creative process: Past, present, and future. *Creativity Research Journal*, 13, 295–308.
Mace, M. A., & Ward, T. (2002). Modeling the creative process: A grounded theory analysis of creativity in the domain of art making. *The Creativity Research Journal*, 14, 179–192.
Martin, B., & Hanington, B. (2012). *Universal methods of design*. Rockport Publishers.
McDowell, A. (2019). Storytelling shapes the future. *Journal of Futures Studies*, 23(3), 105–112.
Mumford, M. D., Reiter-Palmon, R., & Redmond, M. R. (1994). Problem construction and cognition: Applying problem representations in ill-defined domains. In M. A. Runco (Ed.), *Problem finding, problem solving, and creativity* (pp. 3–39). Ablex.
Osborn, A. F. (1963). *Applied imagination* (3rd ed.). Scribners.
Pantopicon. (2022, June 2). https://pantopicon.be
Runco, M. A. (1997). *The creativity research handbook*. Hampton Press.
Simon, H. A. (1996). *The sciences of the artificial* (3rd ed.). The MIT Press.
Stickdorn, M., Hormess, M., & Lawrence, A. (2018). *This is service design methods: A companion to this is service design doing*. O'Reilly Media.
Torrance, E. P. (1962). *Guiding creative talent*. Prentice-Hall.
Treffinger, D. J. (1995). Creative problem solving: Overview and educational implications. *Educational Psychology Review*, 7, 301–312.
Wakslak, C. J., & others (2006). Seeing the forest when entry is unlikely: Probability and the mental representation of events. *Journal of Experimental Psychology: General*, 135(4), 641–653.
Wallas, G. (1926). *The art of thought*. Harcourt, Brace and Company.

7 AI Empathy in the Metaverses

Chiju Chao

BACKGROUND

HUMAN-MACHINE SOCIETY

In recent years, there has been a continual intersection and development of research in artificial intelligence (AI) and human-computer interaction (HCI). As AI technology advances through data feedback and technological progress, it concurrently fuels innovation in HCI. Notably, at the technical level, tools such as voice interfaces, computer vision, machine learning, and recommendation algorithms have emerged as powerful instruments supporting HCI. On the application front, virtual agents, social robots, and user service robots serve as exemplary instances of well-integrated approaches within the field of AI (Ma, 2018).

Examining the developmental history of AI, it becomes apparent that this technology has a relatively brief history, with breakthroughs emerging only in the 21st century. Although AI was proposed by researchers as early as the 1940s and achieved some success through applications such as chess programs, it remained in the computational reasoning stage. During this period, AI essentially replicated human logical capabilities in computational reasoning and was confined primarily to laboratory research, lacking practical applications in business.

In 2006, Jeffrey Hilton and colleagues introduced the concept of "deep learning," representing a significant advancement in machine learning and propelling AI into a new phase of development. With the evolution of deep learning technology, AI products have progressively aligned with people's understanding of "intelligence." They can not only adeptly handle various complex tasks but also engage in natural conversations and emotional interactions with humans (Gu, 2018).

We find ourselves amidst the fourth Industrial Revolution, an era dominated by technological advancements, particularly in AI. This technological surge has given birth to a diverse range of products, profoundly altering the fabric of our lives. Indeed, throughout history, societal progress has undergone transformative shifts in sync with the evolution of industrial revolutions (Xu et al., 2018)..

Commencing with the first Industrial Revolution in the 1760s, known as the machine age, the invention of machines aimed to liberate a significant portion of the workforce. The second Industrial Revolution in the 19th century, often referred to as the age of electricity, witnessed the widespread use of generators illuminating the night and introducing household appliances into every home, thereby reshaping people's daily routines. The latter half of the 20th century marked the advent of the third Industrial Revolution, the age of technology, characterized by the widespread adoption

DOI: 10.1201/9781032693606-10

of computers and the internet, reducing the distances in human communication. This period had far-reaching effects on social structures, production relations, ethics, and morality, concurrently widening socioeconomic disparities (Schwab, 2017).

The ongoing fourth Industrial Revolution is characterized by comprehensive technological breakthroughs, encompassing AI, new energy sources, biotechnology, and nuclear technology. Focusing on the development of AI technology, it has given rise to various intelligent products influencing both industrial processes and daily life. For instance, the ubiquitous robotic vacuum cleaners, equipped with autonomous navigation and recharging capabilities, tirelessly await commands for floor cleaning, becoming an indispensable "companion" for many households.

Our society is gradually transitioning toward a harmonious coexistence of humans and machines in response to these technological developments. People's lives are increasingly intertwined with a myriad of intelligent products, ranging from those performing mechanized tasks (such as robotic vacuum cleaners and logistics transport robots) to providing companionship for entertainment (such as smart speakers and virtual pets), or serving as efficient assistants to humans (such as intelligent in-car navigation systems). In this evolving human-machine society, designers and researchers should actively contemplate how to construct higher-experience, high-value, and ethically sound human-machine relationships, offering valuable insights for the future design of intelligent products.

Perspective on Human-Machine Collaboration

In the realm of intelligent products, the relationship between humans and machines is termed "collaborative interaction." This involves individuals with shared goals engaging in interactive relationships, collectively undertaking activities to achieve common objectives (Sidner et al., 2005). In the current era of advanced AI technology, the interaction between humans and intelligent products mirrors collaborative dynamics seen in HCIs. This collaborative interaction is crucial due to the complementary capabilities of humans and AI, offering substantial advantages for team efficiency in professional and daily contexts, addressing various aspects of team decision-making.

AI excels in addressing complex problems requiring mathematical analysis, while humans contribute creative and intuitive approaches, especially in uncertain and ambiguous domains (Jarrahi, 2018) (Damm, 2012). For instance, AI-driven natural language processing in voice interaction facilitates seamless human-machine communication, allowing users to query information directly, bypassing the time-consuming process of conventional internet searches. In the field of HCI, researchers move beyond viewing AI as a mere tool. Instead, they strategize ways to enhance collaborative interactions between AI systems and humans (Cho & Rader, 2020).

However, many researchers in the field of HCI argue that the relationship between humans and intelligent products extends beyond mere collaborative interaction to a symbiotic relationship—a mutually beneficial association (Cho & Rader, 2020; Licklider, 1960). As the initial goal of AI was to make machines gradually approach human capabilities, some researchers in the HCI domain view AI as "quasi-human" and define the relationship between AI and humans in terms of symbiosis. They note that our human-machine relationships are transitioning from a non-symbiotic present toward an anticipated symbiotic future (Licklider, 1960).

In the human-machine society, for AI to seamlessly integrate into human society and become an excellent collaborator, it needs to be designed and developed to be a reliable and trustworthy partner within human teams (Paschkewitz & Patt, 2020). Therefore, besides focusing on the benefits that AI brings to interactive collaboration, considerations should extend to the "emotional intelligence" of AI. Relevant intelligent technologies include semantic recognition, emotion recognition, etc.—enabling AI to autonomously understand human behavior, thoughts, emotions, provide appropriate feedback, and continuously learn to adapt to human partners, much like humans adapt to the behaviors of human teammates (Chakraborti et al., 2017). Such technologies imbue AI with emotional attributes, making it more engaging for users and more readily accepted within user teams, a design trend already evident in many contemporary AI products.

The direction of design is expanding beyond interaction functionality, aesthetic appeal, and material attributes to incorporate intelligent features such as dialogue patterns, behavioral expressions, and emotional communication. The design perspective is no longer solely about meeting user needs or enhancing user experience but also about constructing a social integration between humans and products. Numerous designers have proposed research and insights in this regard, leveraging social psychology, attempting to design "character" prototypes for intelligent products, such as police-type, butler-type, friend-type (Lin, 2017) Alternatively, starting from personality dynamics, personalized character dimensions are designed for products (LuxAI, 2022).

As AI technology and products gradually permeate our lives, forming the trend of a human-machine society, computer technology researchers are rapidly advancing the technical capabilities of AI, bringing forth various challenges: How should social relationships between humans and intelligent products be constructed? How can collaborative interaction and emotional interaction be balanced in different scenarios? Design researchers should leverage their strengths to conduct design research, addressing these questions, keeping pace with technological developments, or even surpassing the speed of technological advancement to provide the most socially valuable guidance for intelligent product design for human users.

METAVERSE

What Is Metaverse

In 2021, the concept of the metaverse gained prominence, also referred to as the "metaverse era." The term "metaverse" is relatively new and is commonly used to describe a virtual, digitized world. It goes beyond a typical digital world and represents a parallel and independent virtual space alongside the real world—a digitally mapped online realm that is becoming increasingly realistic. Within the metaverse, people can engage in various activities, including socializing, working, learning, and entertainment. The metaverse is a convergence of technologies such as virtual reality (VR), augmented reality (AR), AI, aiming to create a digitized environment similar to or even richer than the real world.

Many leading researchers have proposed concepts, definitions, and scopes for the metaverse. For example, the metaverse must provide "unprecedented interoperability"—users must be able to carry their avatars and goods from one part of the metaverse to another, regardless of who operates that particular part of the metaverse (Ball, 2022). The metaverse includes at least the following elements: identity, friends, immersion, low latency, diversity, ubiquity, economic systems, and civilization (Roblox, 2024). The seven layers of the metaverse construction are: experience, discovery, creator economy, spatial computing, decentralization, human-computer interaction, and infrastructure (Duan et al., 2021). We believe all of these definitions are correct, as the metaverse is an enormous concept involving the bidirectional flow between physical and virtual realms, making it impossible to define precisely "what it is."

In the metaverse, information from physical spaces is flowing into virtual spaces, leading to the development of bidirectional communication between virtual and physical spaces (Zhuang et al., 2018). Through the integration of information, the boundaries between the real and virtual worlds are broken, seamlessly connecting the two realms. This not only enhances the realism of the virtual world but also injects new vitality into the physical world (Gushima & Nakajima, 2017). In the metaverse, users can create, interact, and share content, facilitating connections across geographical, cultural, and linguistic boundaries. The development of the metaverse is believed to have profound impacts on social, business, educational, and other domains. However, it is important to note that the concept of the metaverse is still evolving, and its definition and applications may continuously evolve with technological advancements in the future.

In the metaverse, AI plays a pivotal role as a key technology underpinning the entire virtual environment. For instance, AI can be utilized to create virtual assistants and NPCs (non-player characters), offering assistance, interaction, and entertainment. AI customizes and optimizes the metaverse experience based on user behavior, interests, and preferences, engaging in natural interactions with users through natural language processing technology. The advancement of AI technology in the metaverse aims to make virtual experiences more intelligent, immersive, and provide users with a more personalized and enriched digital environment. However, ethical and privacy issues associated with this development also demand close attention.

In this trend, the interactive collaborative scenarios between humans and machines may transition to the virtual world. AI assistants can offer support in the metaverse, aiding users in tasks, providing real-time suggestions, and optimizing workflows. They can also collaborate with users in creating and editing content within the virtual environment. For designers, one of the key challenges is ensuring the smooth and efficient collaboration between humans and AI while also taking into consideration security and privacy protection.

3D Design Tools in the Metaverses

In the traditional physical world, collaborative design typically involves using flat tools such as paper, pens, sticky notes, and whiteboards as mediums to initiate design activities. Correspondingly, collaborative discussion tools are also designed in a flat format. However, when engaging in collaboration in today's VR environment, we recognize the need to explore new concepts for the interaction design of relevant tools. There are two primary reasons behind this need.

First, in VR environments, people may not necessarily excel at handling traditional flat materials. Traditional paper and whiteboards may not fully replicate their effectiveness in a physical environment within the virtual world. Therefore, we need to reconsider how design collaboration takes place in virtual space to better align with users' interactive capabilities in VR environments.

Second, virtual environments offer the advantage of achieving spatial changes at a lower cost, providing us with an opportunity to experiment with innovative design tools. In VR, users can more easily alter their work environment, rearrange elements without the need for reprinting or resetting physical tools. Thus, we have the opportunity to explore more three-dimensional, concept-based tool designs.

To better adapt to this new environment, we are committed to three-dimensionalizing the collaborative design process and design tools, allowing users to engage in discussions and design by manipulating three-dimensional objects within the virtual environment. This approach provides users with a more intuitive, immersive experience, enabling them to interact more naturally with the virtual space and express and share their creativity more flexibly. By introducing three-dimensional design elements, we aim to drive collaborative activities toward more innovative and cooperative directions, thus achieving more productive and satisfying design processes in the virtual environment.

In relevant design research, researchers are dedicated to constructing three-dimensional design tools in virtual space, transforming tool content filled on flat surfaces in the real world into three-dimensional graphics, and implementing corresponding interactions (Chao et al., 2023). This three-dimensional approach has gained widespread user recognition. The three-dimensional design method has significant advantages, as roaming and movement actions are relatively easy in virtual space, whereas fine flat operations can be more challenging. From an interactive experience perspective, this three-dimensional design method is better suited for virtual space, representing an innovative concept with continuous development potential.

Although this three-dimensional approach excels in interactivity, there may be some difficulties in understanding and learning due to the fact that most users have not encountered similar tools. This is a challenge that needs continuous improvement and resolution in subsequent research. To better promote and popularize this three-dimensional interactive design, attention needs to be given to the user learning curve, providing effective guidance and training to help users become familiar with and master the use of this new type of tool.

When undertaking such three-dimensional interactive design, it is essential to consider not only the interactive form but also factors such as spatial layout, rational arrangement of processes, and guided teaching. Design decisions in these aspects will directly impact the user acceptance and effectiveness of the tool. Therefore, in future research and development, we will delve into optimizing these key design elements to enhance the user experience and seamlessly integrate three-dimensional design tools into virtual space.

AI EMPATHY

WHAT IS EMPATHY

The concept of empathy was first introduced by the humanistic founder Carl Rogers and refers to the ability to experience the inner world of others. Empathy, as a psychological concept, has been extensively studied in the discipline of psychology. Currently, psychologists offer two main theoretical explanations for empathy: "cognitive process" and "emotional process." The "cognitive process" posits that empathy requires the subject to activate emotions through imagination and reflection, eliciting a shared emotional experience with the object and prompting corresponding behavioral responses. On the other hand, the "emotional process" suggests that the subject can directly perceive the emotional state of the other, leading to a matching emotional response. While these two theories remain subject to ongoing debate among researchers, they both have substantial experimental evidence and various mature process models. These process models position empathy as a higher-level category encompassing all phenomena with similar mechanisms, such as emotional contagion, sympathy, cognitive empathy, and helping behavior. Currently, nearly all empathic processes rely on these process models for design, demonstrating adaptability and even existing across various species (Preston & de Waal, 2002).

Empathy plays a crucial role in interpersonal relations, where individuals continuously assess each other's situations, emotions, and behaviors. In response, they adjust their own emotional states and express empathic behaviors, exemplifying everyday manifestations of empathy (McQuiggan & Lester, 2007). In fact, empathy often serves as the motivation behind altruistic actions in interpersonal interactions. Psychologists have confirmed that individuals become more willing to help and generous after experiencing empathy, fostering a positive cycle of empathic interactions with others (Singer & Lamm, 2009). Empathy effectively promotes emotional communication and sustains the smooth development of interpersonal relations, playing a pivotal role in societal well-being.

Studies indicate that empathy between individuals in team collaboration has numerous beneficial effects on attitudes and behaviors. Conversely, individuals lacking empathy can contribute to negative impacts during collaborative processes. Therefore, effectively harnessing empathy in team collaboration to build positive empathic interactions can significantly benefit projects. Researchers in the theory of cognitive empathy suggest that empathy facilitates understanding important environmental information about others and predicting their future behaviors (de Vignemont & Singer, 2006). In the context of team collaboration, empathy not only assists in positive emotional communication but also aids individuals in effectively obtaining information about others' environments and predicting their actions.

Beyond fostering positive communication, empathy in team collaboration also influences the quality of reasoning and decision-making. Psychologists affirm the crucial role of human emotions in rational thinking, emphasizing the need for emotional input to guide behavioral decisions (Dalvandi, 2013). Similar to individual cognitive processes, teams undergo cognitive processes encompassing learning, planning, reasoning, decision-making, problem-solving, memory, design, and situation evaluation. Consequently, teams, like individuals, may make suboptimal decisions or engage in less-than-ideal behaviors (Cooke et al., 2013). Empathy, when teams make erroneous decisions, facilitates mutual understanding among members. It aids in information exchange, identifies the root causes of errors, and resolves conflicts among members to facilitate collaborative decision-making (Stephan & Finlay, 1999).

Given the core role of emotional factors in human relationships, they should play an equally crucial role in intelligent products (McQuiggan & Lester, 2007). Motivation, empathy, and social skills contribute significantly to mutual understanding, knowledge exchange, and seamless cooperation among team members, holding great significance (Luca & Tarricone, 2001). In the realm of human-machine collaboration, incorporating emotional mechanisms is essential for enhancing collaboration efficiency (Dalvandi, 2013). Research confirms that the personality of AI is a key factor influencing people's empathy toward human-machine collaboration and their satisfaction with the experience (Afzal et al., 2019).

Existing research has demonstrated that intelligent products capable of generating empathetic responses are more likable and trustworthy than agents lacking empathy capabilities (Brave et al., 2005). If the core system of empathy is integrated into intelligent products, leveraging deep learning in computers, these products can potentially undergo self-learning through human-machine interactions, thereby developing empathy systems similar to humans (Asada, 2015). Such empathetic responses go beyond endowing robots with perceptual capabilities; they also involve assessing the emotions of human communication partners to make interactions more natural (Hegel et al., 2006). Empathy is a two-way process, emerging from the interaction between two entities. Therefore, when empathy serves as a design principle in the development of intelligent products, attention should be given not only to the design of the product's empathetic interaction but also to the evolving emotional experiences of humans throughout the empathetic process.

Today, the fields of design, psychology, and computer science are all engaged in discussions about empathy in intelligent products. Design, for instance, explores empathetic interaction design from the perspective of emotional design. Psychology endeavors to provide algorithmic models for empathetic computation based on human empathy theory models, while computer science, coupled with AI technology, offers implementable capabilities for empathetic simulation. In this section, we discuss these three directions. First, we provide an overview of research on empathy design between humans and intelligent products in various domains. Second, we clarify disciplinary boundaries to help orient future design research.

Empathetic Design Perspective

Thanks to the capabilities of AI, contemporary products can now engage in affective computing by combining technologies such as computer vision and semantic analysis to analyze and assess users' emotional states. Imagine if intelligent products could provide appropriate interactive feedback based on users' emotional states, creating a sense of empathy. In doing so, these products simulate the empathetic abilities inherent in human beings.

In reality, humans tend to anthropomorphize objects and project emotions onto them. Even when faced with products devoid of any intelligent capabilities or personification attributes, humans still project emotions onto them. Psychologists argue that anthropomorphizing objects can bring great joy, delight, and satisfaction to users (Norman, 2004). German aesthetician Friedrich Theodor Vischer also mentioned the concept of "aesthetic symbolism," where "a person projects himself into objects in the natural world," resulting in the "humanization of the object." Practices like anthropomorphizing objects or attempting to foster empathy with products do not fall under the practical elements of a product. While they may not pragmatically fulfill people's needs, they can deliver high-value emotional experiences.

The research on product design in this direction was introduced as early as the 1980s and was termed "Affective Design," indicating that, beyond the practical elements of a product, there exists another critical emotional component (Norman, 2004). Human emotions toward products refer to consumers' sustained emotional attachment and feelings toward products through prolonged interaction and usage (Chapman, 2012). In the past, emotional elements were primarily a one-sided emotional imagination by users, with products rarely engaging in emotional interaction and communication with users. Today, with intelligent products capable of discerning user emotions, the integration of emotions between products and users is bound to be more intimate, making affective design more interactive and complex.

When designers adopt "empathy" as a design concept for intelligent products, they can make significant contributions to the field of intelligent product design. Many scholars believe that the expression of empathy in the HCI process has tremendous social value and significance (Zong & GuangXin, 2016). For example, in a study on intelligent teaching, an intelligent system with empathetic capabilities can carefully consider interpersonal interaction in teaching and provide students with higher-quality support in the classroom. While intelligent agents cannot truly empathize with students, they can improve the learning environment and help meet students' personalized learning needs by exhibiting features of empathetic capabilities through algorithm-generated behaviors (Cooper et al., 2000). Moreover, in the process of human-computer collaboration, empathy contributes to the improvement of collaborative efficiency. For instance, through emotional expressions, intelligent products make it easier for humans to understand the capabilities of the product, determining suitable tasks and learning how to make optimal use of the machine's abilities (Norman, 2004).

When designing the empathic interactive capabilities of intelligent products, we may refer to principles from psychology for theoretical construction. However, their expression of emotions may not necessarily be identical to that of humans, given that people are dealing with created entities rather than genuine life forms. This presents a complex challenge for designers—how to design and manufacture intelligent products, and how to enable them to engage in empathic communication with humans. It may not replicate human emotions but rather be an emotion system tailored to meet the needs of machines (Norman, 2004). Nevertheless, it can still be termed as emotion as it is customized to fulfill the requirements of robots. Simultaneously, ethical and moral considerations are crucial, especially when such products have the potential to evoke strong emotional attachments in humans or even replace emotional interactions between individuals. Ethical and moral issues become particularly significant in such scenarios.

Implementation of Empathy

Social psychologists describe the process of empathy as a cognitive process that involves the intersection of emotion and reason, consisting of three parts: first, the empathizer's consideration of themselves, such as their purpose, intentions, emotional state, and the current situation; second, the evaluation of antecedents, external circumstances, the empathic object, and self; and finally, the outcome of empathy, typically manifested in language, actions, behavior, and expressions, such as expressions of goodwill and caring behavior, which are the products of evaluation (Leite et al., 2013; McQuiggan & Lester, 2007). Therefore, research on empathy does not solely focus on emotional processes; emotional thinking and rational thinking have some intersection and are not mutually exclusive (Jack et al., 2013).

In the field of HCI within computer science, researchers have incorporated empathy models from psychology to develop various types of AI capabilities for human-computer empathy issues. They include factors influencing empathic responses, such as familiarity, similarity, learning ability, past experiences, and emotional salience, in mathematical models (Rodrigues et al., 2009).

The simulation of empathy in intelligent products originated from affective computing technology. In 1986, Professor Minsky at MIT first introduced the problem of emotion recognition in intelligent machines in "The Society of Mind" (Minsky, 1988). In 1997, Professor Picard proposed the framework of affective computing in the book "Affective Computing," stating that affective computing is a computational science concerning human emotion generation, emotion recognition, emotion representation, and the measurement of factors influencing emotions. This formalized the birth of affective computing science and propelled developments in areas such as robotics, machine vision, HCI design, speech recognition, and text analysis (Picard, 1997).

The current mainstream research in affective computing technology primarily divides into two categories: Emotional Analysis and Sentiment Analysis (Rao et al., 2018). Emotional Analysis, with Professor Picard from MIT as a prominent representative, focuses on the collection of emotional information, emotional recognition analysis, emotional understanding cognition, and the expression of emotional information. It classifies emotions based on features and measures their intensity (Taylor et al., 2020). Sentiment Analysis, represented by Professor Liu from UIUC, suggests analyzing the opinions, tendencies, attitudes, and stances related to a target to determine its sentiment. This branch of affective computing considers not only users, organizations, and products but also focuses on issues, events, and topics (Liu, 2012).

As affective computing technology matures, the perception and processing capabilities of computational systems continue to improve, making it possible to handle social and emotional cues and interact naturally with people. Artificial Empathy computing is a new paradigm emerging from this environment, referring to the ability of computational systems to understand and respond to human thoughts, feelings, and emotions. Numerous research studies have explored different computational models for artificial empathy, which may share similar or distinct theoretical foundations. The differences in these models arise from the application scenarios and purposes (Yalçın & DiPaola, 2020). Research indicates that for designing emotionally intelligent products in a specific scenario, an accurate assessment of the social context and an evaluation of the appropriate type of empathic behavior are necessary (McQuiggan & Lester, 2007).

As a designer, artificial empathy computing technology in the field of computer science not only facilitates the realization of envisioned empathic interactive capabilities but also allows designers to contribute to the design elements focused on expressions, such as the design of empathic responses in intelligent products and the design of modulating factors. Although computer technology enables intelligent products to recognize and analyze emotions based on scene perception, emotional cues, body language, or tone of voice, and calculate the correlation between emotion types and modulating factors (Lim & Okuno, 2015), the post-computational behavioral feedback and the factors influencing the computation need to align with user needs. For instance, the computer may calculate a potential empathetic emotion like "sadness," but depending on the product type, scenario, and user, the object may require comfort, a solution, companionship, a warm hug, soothing music, and various other modulating factors and expressions of empathic responses, all of which are subjects for research and exploration in this field by designers.

CASE STUDY

Our team conducted an exploratory design study on the theme of AI empathy in the metaverse. We aimed to find inspiration for an in-depth investigation into this topic and seamlessly integrate various possible research concepts. Through active experimentation in practical tests, we hope to discover unexpected user reactions that could drive innovation in this field.

In terms of AI emotional resonance, we focused on the user experience in the metaverse environment, attempting to realize AI's perception and response to user emotions in virtual space. We consolidated various possible research concepts, ranging from virtual agents expressing emotions to social experiences involving emotional resonance. By combining these concepts, we aim to explore the multifaceted nature of AI emotional resonance in the metaverse, gaining a more comprehensive understanding and practical application in this exciting domain. We adopted a method of practical testing, seeking to capture users' authentic reactions during their interactions with AI emotional resonance systems through direct engagement.

In our research, we introduced personality traits to intelligent assistants. We initially experimented with two dimensional attributes: rationality versus emotionality and proactivity versus passivity. The rational assistant provides users with necessary information content, presenting only the information required by the user. On the other hand, the emotional assistant expresses different emotional reactions based on the user's state, using pre-set emotional tone words. The proactive assistant automatically triggers interactive behaviors, presenting information content and initiating automatic voice playback. The passive assistant requires users to ask questions verbally before displaying content and does not engage in voice playback unless prompted by the user.

During the co-creation discussion phase, we used three design tools—Future Signs STEEP, and Future Triangle—to guide users through the co-creation process. First, we collected future signs related to the design theme from users in the form of images and keywords. This data was recorded on a digital platform and displayed on the walls of the VR space. In the second step, users collaboratively discussed and selected eight future signs that aligned with a consistent concept. In the third step, once the eight future signs were selected, the keywords associated with the five elements of society, technology, environment, economy, and politics implied by these future signs were displayed on five pillars in the scene. Users could then choose, delete, or edit these keywords.

In the interaction design, we use gaze interaction, voice interaction, and gesture interaction. There are different interaction behaviors in different scenarios (Table 7.1):

We designed three user scenarios: Character Creation Space, Co-creation Discussion Space, and Results Presentation Space. The scene models were created using C4D and prototyped on the Unity3D platform (Figure 7.1).

Upon entering the virtual environment, users first enter the Character Creation Space, which is primarily responsible for generating user roles and selecting AI assistant personalities. Once confirmed, users will be transported to the Co-creation Discussion Space, where they engage in collaborative discussions involving the selection of future resource libraries, editing STEEP keywords,

TABLE 7.1
Multimodal Interaction Design in Three Scenarios

Scenarios	Future Signs (Hiltunen, 2008)	STEEP (Szigeti et al., 2011)	Future Triangle (Fergnani, 2020)
Gaze interaction	By gazing, some keywords about the target being gazed at can be listed around the target.	After focusing on a certain keyword, the intelligent assistant will list the images and information related to it around the target.	By gazing at the keywords, the intelligent assistant displays the relevant information that has been discussed around.
Voice interaction		By analyzing what the user is currently saying through NLP technology, the user can get advice from the intelligent assistant at any time.	The intelligent assistant deduces where the user's input is most likely to be posted and gives the user a hint.
Gesture interaction	Select a target and place its details in the text box in the head of the intelligent assistant.	When the gesture points to a keyword, the intelligent assistant conducts an internet search and displays more detailed information to the user.	

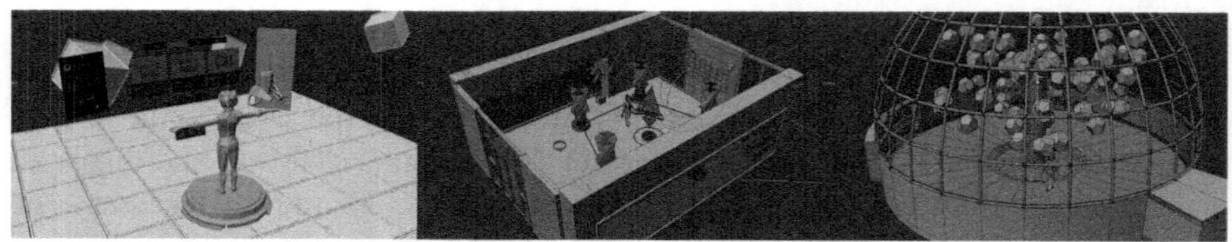

FIGURE 7.1 Modeling Prototypes for 3 Scenarios in Unity

and conducting future triangle analyses. Finally, users will enter the Results Presentation Space, which documents the outcomes of all groups that participated in the collaborative session, presented in the form of a knowledge tree. We developed the Synneure usage process on the Unity3D platform and utilized the capabilities of the Baidu AI open platform for speech recognition input, AI. natural interaction, and natural language analysis.

In this platform, we constructed a three-dimensional design tool in the virtual space, allowing tools that are typically filled out in a flat manner in the real world to be represented and interacted with in three-dimensional graphics. This approach has been well-received by the majority of users. Due to the ease of roaming movements in virtual space, fine flat operations, on the contrary, are more challenging. Therefore, from the perspective of user interaction, this three-dimensional approach is more suitable for virtual space and is a concept that can be further developed. However, since most people have not used similar tools before, there is a certain difficulty in understanding and learning, which is something we need to improve in the future.

Additionally, we found that there are more stage changes in team collaboration in the virtual environment compared to the real environment. These changes stem from the user's process of becoming familiar with the virtual environment and the transition from needing to not needing information retrieval. Therefore, the responsibilities of the AI assistant in the virtual environment also need to adapt to these changes.

In terms of emotional experiences, we found that once users experience communication failures with the AI assistant, they often feel distrust, leading to a poor experience. Emotional expression can play a relatively effective role in mitigating this, making the AI assistant more humane, and encouraging users to give it another chance. However, this also brings certain ethical concerns. One participant expressed the view that such personality traits might influence his judgment, allowing the AI assistant to shape the collaborative process, which he considered negative for human teams.

In the metaverse virtual world, users can have more interactive operations with the AI assistant. These interaction designs are easier to implement compared to products in the real world. For this reason, we found that most users, upon entering the virtual world, attempt various interactive actions with the AI assistant. However, since this study did not focus on dynamic interaction design, users were unable to receive feedback. This led us to identify a valuable design direction for the emotional interactive expression of AI assistants in the virtual world.

CULTURE AND SOCIAL RESPONSIBILITY

The initial goal of AI technology was to make computers approach human capabilities infinitely. When this goal becomes achievable in the future, humans may perceive AI as a "quasi-human." Researchers predict that in the future, the human-machine relationship will evolve from a subject-object relationship into a new type of intersubjective relationship—interpersonal intersubjectivity (Cheng, 2019). As AI can autonomously perceive the world and choose actions independently, people might see it as a subject with emotions, intelligence, and capabilities beyond those of humans. Consequently, humans are beginning to establish ethical norms specifically for AI (Hu & Xiang, 2020).

The ethical and moral discussions regarding AI products are complex. Despite the construction of numerous "man-machine symbiotic societies" scenarios in various literary works, such as science fiction novels and movies, which allows the public to envision corresponding social landscapes, philosophers, designers, psychologists, and other researchers still find it challenging to define the ethics of AI. This complexity primarily arises from two aspects: the first being the complexity of AI facing human moral life, and the second being how humans attribute moral behavior and capabilities to AI, a complexity that designers must address (Hu & Xiang, 2020).

This chapter discusses AI technology with an optimistic perspective, but concerns about AI are not unfounded. The concept of "symbiosis between humans and machines" as a new cooperative decision-making relationship raises valid worries. As the autonomy of AI entities strengthens, and their cooperation with humans deepens, it brings forth certain risks (Jianhua, 2018), such as new ethical, political issues, data misuse, privacy risks, and emotional concerns. With the advent of a risk society, humanity needs to seek a "second-tier decision support system" to face a society with high complexity and uncertainty (Damm, 2012). This may give rise to endless and challenging issues, potentially leading to the collapse of the entire system.

For designers, it is crucial to approach the design of AI entities from the perspectives of moral ethics and cultural responsibility. One critical issue that many researchers focus on is whether AI entities should undergo anthropomorphic design. While a highly anthropomorphic appearance may make AI more human-like, causing humans to gradually consider them as peers, the problem of consciousness is a longstanding challenge for these physical entities. Such issues are essential considerations for designing truly empathetic AI entities. Other design aspects, such as facial expressions, body language, and language, can influence people's moral perceptions of AI entities. Collaboration patterns and data usage also play roles in shaping people's moral value judgments, all of which are aspects this research needs to consider.

REFERENCES

Afzal, S., Dempsey, B., D'Helon, C., Mukhi, N., Pribic, M., Sickler, A., Strong, P., Vanchiswar, M., & Wilde, L. (2019). The personality of AI systems in education: Experiences with the Watson Tutor, a one-on-one virtual tutoring system. *Childhood Education*, *95*(1), 44–52. https://doi.org/10.1080/00094056.2019.1565809

Asada, M. (2015). Towards artificial empathy. *International Journal of Social Robotics*, *7*(1), 19–33. https://doi.org/10.1007/s12369-014-0253-z

Ball, M. (2022). *The Metaverse: And How It Will Revolutionize Everything.* Liveright Publishing.

Brave, S., Nass, C., & Hutchinson, K. (2005). Computers that care: Investigating the effects of orientation of emotion exhibited by an embodied computer agent. *International Journal of Human-Computer Studies, 62*(2), 161–178. https://doi.org/10.1016/j.ijhcs.2004.11.002

Chakraborti, T., Kambhampati, S., Scheutz, M., & Zhang, Y. (2017). *AI Challenges in Human-Robot Cognitive Teaming* (arXiv:1707.04775). http://arxiv.org/abs/1707.04775

Chao, C., Wang, Q., Wu, H., & Fu, Z. (2023). Synneure: Intelligent Human-Machine Teamwork in Virtual Space. In P.-L. P. Rau (Ed.), *Cross-Cultural Design* (Vol. 14023, pp. 349–361). Springer Nature Switzerland. https://doi.org/10.1007/978-3-031-35939-2_25

Cheng G. (2019). From A Man-Machine Relationship to Inter Human Relations: Definition and Strategy of Artificial Intelligence. *Journal of Dialectics of Nature, 41*(01), 9–14. https://doi.org/10.15994/j.1000-0763.2019.01.002

Chapman, J. (2012). *Emotionally Durable Design* (1st ed.). Routledge. https://doi.org/10.4324/9781849771092

Cho, J., & Rader, E. (2020). The role of conversational grounding in supporting symbiosis between people and digital assistants. *ACM on Human-Computer Interaction, 4(CSCW1)*, 1–28. https://doi.org/10.1145/3392838

Cooke, N. J., Gorman, J. C., Myers, C. W., & Duran, J. L. (2013). Interactive team cognition. *Cognitive Science, 37*(2), 255–285. https://doi.org/10.1111/cogs.12009

Cooper, B., Brna, P., & Martins, A. (2000). Effective Affective in Intelligent Systems – Building on Evidence of Empathy in Teaching and Learning. In A. Paiva (Ed.), *Affective Interactions* (Vol. 1814, pp. 21–34). Springer Berlin Heidelberg. https://doi.org/10.1007/10720296_3

Dalvandi, B. (2013). *A model of empathy for artificial agent teamwork.* [Doctoral dissertation, University of Northern British Columbia]. https://doi.org/10.24124/2013/bpgub878

Damm, L. (2012). Moral machines: Teaching robots right from wrong. *Philosophical Psychology, 25*(1), 149–153. https://doi.org/10.1080/09515089.2011.583029

de Vignemont, F., & Singer, T. (2006). The empathic brain: How, when and why? *Trends in Cognitive Sciences, 10*(10), 435–441. https://doi.org/10.1016/j.tics.2006.08.008

Duan, H., Li, J., Fan, S., Lin, Z., Wu, X., & Cai, W. (2021). Metaverse for Social Good: A University Campus Prototype. In Proceedings of the 29th ACM International Conference on Multimedia (pp. 153–161). https://doi.org/10.1145/3474085.3479238

Fergnani, A. (2020). Futures Triangle 2.0: Integrating the Futures Triangle with Scenario Planning. *Foresight, 22*(2), 178–188. https://doi.org/10.1108/FS-10-2019-0092

Gu, J. (2018). Artificial Intelligence: A Brief History of Development, Technical Case Studies, and Business Applications. *Sci-Tech Innovations and Brands*, 7, 4–4.

Gushima, K., & Nakajima, T. (2017). A design space for virtuality-introduced internet of things. *Future Internet, 9*(4), 60. https://doi.org/10.3390/fi9040060

Hegel, F., Spexard, T., Wrede, B., Horstmann, G., & Vogt, T. (2006). Playing a Different Imitation Game: Interaction with an Empathic Android Robot. *2006 6th IEEE-RAS International Conference on Humanoid Robots*, 56–61. https://doi.org/10.1109/ICHR.2006.321363

Hiltunen, E. (2008). The future sign and its three dimensions. *Futures, 40*(3), 247–260. https://doi.org/10.1016/j.futures.2007.08.021

Hu, S. & Xiang, Y. (2020). The Complexity of AI Moral Judgment and its Solution——From the Perspective of "Man-computer Symbiosis". *Journal of Jiangsu University (Social Sciences Edition), 22*(04), 16–28. https://doi.org/10.13317/j.cnki.jdskxb.2020.036

Jack, A. I., Dawson, A. J., Begany, K. L., Leckie, R. L., Barry, K. P., Ciccia, A. H., & Snyder, A. Z. (2013). fMRI reveals reciprocal inhibition between social and physical cognitive domains. *NeuroImage, 66*, 385–401. https://doi.org/10.1016/j.neuroimage.2012.10.061

Jarrahi, M. H. (2018). Artificial intelligence and the future of work: Human-AI symbiosis in organizational decision making. *Business Horizons*, *61*(4), 577–586. https://doi.org/10.1016/j.bushor.2018.03.007

Jianhua, M. (2018). Understanding and theory: The optimistic attitude and pessimistic attitude in the discussion of artificial intelligence. *Journal of Dialectics of Nature*, *40*(4), 1–8. https://doi.org/10.15994/j.1000-0763.2018.04.001

Leite, I., Pereira, A., Mascarenhas, S., Martinho, C., Prada, R., & Paiva, A. (2013). The influence of empathy in human–robot relations. *International Journal of Human-Computer Studies*, *71*(3), 250–260. https://doi.org/10.1016/j.ijhcs.2012.09.005

Licklider, J. C. R. (1960). Man-computer symbiosis. *IRE Transactions on Human Factors in Electronics*, *HFE-1*(1), 4–11. https://doi.org/10.1109/THFE2.1960.4503259

Lim, A., & Okuno, H. G. (2015). A recipe for empathy. *International Journal of Social Robotics*, *7*(1), 35–49. https://doi.org/10.1007/s12369-014-0262-y

Liu, B. (2012). *Sentiment Analysis and Opinion Mining*. Springer International Publishing. https://doi.org/10.1007/978-3-031-02145-9

Luca, J., & Tarricone, P. (2001). Does emotional intelligence affect successful teamwork? *18th Annual Conference of the Australasian Society for Computers in Learning in Tertiary Education*, 367–376. https://ro.ecu.edu.au/ecuworks/4834

LuxAI. (2022). *QT Robot*. Retrieved 31 October, 2022, from https://luxai.com/

Ma, X. (2018). Towards Human-Engaged AI.27th International Joint Conference on Artificial Intelligence, 5682–5686. https://doi.org/10.24963/ijcai.2018/809

McQuiggan, S. W., & Lester, J. C. (2007). Modeling and evaluating empathy in embodied companion agents. *International Journal of Human-Computer Studies*, *65*(4), 348–360. https://doi.org/10.1016/j.ijhcs.2006.11.015

Minsky, M. (1988). *The Society of Mind*. Simon & Schuster. https://discover.lib.tsinghua.edu.cn/entrance/searchEntrance/resourceDetail?id=86THU_ALMA_US21317328230003966&search_scope=default_scope&search=The%20society%20of%20mind&title=The%20Society%20of%20mind%20%2F&version=2&frbrgroupid=518346187&context=L&adaptor=Local%20Search%20Engine&qInclude=facet_frbrgroupid,exact,518346187%7C,%7C&query=any,contains,The%20society%20of%20mind&isFrbr=true

Norman, D. A. (2004). *Emotional Design: Why We Love (or Hate) Everyday Things*. Basic Books. https://discover.lib.tsinghua.edu.cn/entrance/searchEntrance/resourceDetail?id=86THU_ALMA_CN21327758480003966&search_scope=default_scope&search=%E6%83%85%E6%84%9F%E5%8C%96%E8%AE%BE%E8%AE%A1&title=%E8%AE%BE%E8%AE%A1%E5%BF%83%E7%90%86%E5%AD%A6.%203.%20%E6%83%85%E6%84%9F%E5%8C%96%E8%AE%BE%E8%AE%A1%20%3D%20Emotional%20design&version=&frbrgroupid=553488162&context=L&adaptor=Local%20Search%20Engine&query=any,contains,%E6%83%85%E6%84%9F%E5%8C%96%E8%AE%BE%E8%AE%A1&isFrbr=true

Paschkewitz, J., & Patt, D. (2020). Can AI make your job more interesting? *Issues in Science and Technology*, *37*(1), 74–78.

Picard, R. W. (1997). *Affective Computing*. MIT Press. https://discover.lib.tsinghua.edu.cn/entrance/searchEntrance/resourceDetail?id=86THU_ALMA_US51466979820003966&search_scope=default_scope&search=affective%20computing&title=Affective%20computing%20%2F&version=2&frbrgroupid=517949493&context=L&adaptor=Local%20Search%20Engine&query=any,contains,affective%20computing&isFrbr=true

Preston, S. D., & de Waal, F. B. M. (2002). Empathy: Its ultimate and proximate bases. *Behavioral and Brain Sciences*, *25*(1), 1–20. https://doi.org/10.1017/S0140525X02000018

Rao, Y., Wang, Y.-M., Wu, L.-W., & Feng C. (2018). Research Progress on Emotional Computation Technology Based on Semantic Analysis. *Journal of Software*, *29*(8), 2397–2426.

Roblox. (2024). Retrieved August 27, 2024, from https://www.roblox.com

Rodrigues, S. H., Mascarenhas, S. F., Dias, J., & Paiva, A. (2009). "I can feel it too!": Emergent empathic reactions between synthetic characters. *2009 3rd International Conference on Affective Computing and Intelligent Interaction and Workshops*, 1–7. https://doi.org/10.1109/ACII.2009.5349570

Schwab, K. (2017). *The Fourth Industrial Revolution*. Crown.

Lin, S. (2017). *Design of Intelligent Product*. Publishing House of Electronics Industry. https://tsinghua-primo.hosted.exlibrisgroup.com.cn/permalink/f/1secrdm/86THU_ALMA_CN21465934010003966

Sidner, C. L., Lee, C., Kidd, C. D., Lesh, N., & Rich, C. (2005). Explorations in engagement for humans and robots. *Artificial Intelligence*, *166*(1), 140–164. https://doi.org/10.1016/j.artint.2005.03.005

Singer, T., & Lamm, C. (2009). The social neuroscience of empathy. *Annals of the New York Academy of Sciences*, *1156*(1), 81–96. https://doi.org/10.1111/j.1749-6632.2009.04418.x

Stephan, W. G., & Finlay, K. (1999). The role of empathy in improving intergroup relations. *Journal of Social Issues*, *55*(4), 729–743. https://doi.org/10.1111/0022-4537.00144

Szigeti, H., Messaadia, M., Majumdar, A., & Eynard, B. (2011). STEEP analysis as a tool for building technology roadmaps. *Internationale challenges e-2011 conference. 2011*, 26–28.

Taylor, S., Jaques, N., Nosakhare, E., Sano, A., & Picard, R. (2020). Personalized multitask learning for predicting Tomorrow's mood, stress, and health. *IEEE Transactions on Affective Computing*, *11*(2), 200–213. https://doi.org/10.1109/TAFFC.2017.2784832

Xu, M., David, J. M., & Kim, S. H. (2018). The Fourth Industrial Revolution: Opportunities and Challenges. *International Journal of Financial Research*, *9*(2), 90. https://doi.org/10.5430/ijfr.v9n2p90

Yalçın, ÖN., & DiPaola, S. (2020). Modeling empathy: Building a link between affective and cognitive processes. *Artificial Intelligence Review*, *53*(4), 2983–3006. https://doi.org/10.1007/s10462-019-09753-0

Zhuang, C., Liu, J., & Xiong, H. (2018). Digital twin-based smart production management and control framework for the complex product assembly shop-floor. *The International Journal of Advanced Manufacturing Technology*, *96*(1–4), 1149–1163. https://doi.org/10.1007/s00170-018-1617-6

Zong, Y., & GuangXin, W. (2016). Anthropomorphism: The psychological application in the interaction between human and computer. *Psychology : Techniques and Applications*, *4*(5), 296–305. https://doi.org/10.16842/j.cnki.issn2095-5588.2016.05.007

8 Speculative Artifacts in HCI

Lin Zhu

INTRODUCTION

Speculative design approaches are becoming increasingly common in human-computer interaction (HCI) research, offering new ways to imagine and explore possible technological futures. As HCI expands beyond desktop interfaces to emerging technologies in all life contexts (Dourish, 2004; Harrison et al., 2007), speculative methods such as critical design (Dunne, 1999), design fiction (Sterling, 2005), speculative design (Dunne & Raby, 2013), discursive design (Tharp, 2013), material speculation (Wakkary et al., 2016), and speculative enactments (Elsden et al., 2017) have increased in HCI. Which allows researchers to think more critically and creatively about the social and political implications of technology designs.

Overviews of the ranges of speculative, future-oriented, and fictional works have introduced new epistemological positions in HCI (Mankoff et al., 2013). Critical design utilizes "quasi-functionality" in artifacts, making them appear functional yet unfamiliar and revealing invisible assumptions. Enabling "the invisible and lost in the everyday familiar to become visible" (Malpass, 2016). HCI researchers have not only adopted speculative approaches, but also integrated them into their disciplinary practices to explore topics beyond the focus of product, still leveraging the concept of traditional product design through quasi-functional design artifacts. Speculative design itself can be seen as a key intervention in HCI methodologies – the introduction of a way of thinking and approaching that is more forward-looking and extensible than the study of cognitive behavior and the interaction between individual users and interfaces.

New opportunities have been opened up in the HCI space. We believe it is particularly important to share within HCI's multi-epistemology how artifacts are conceptually framed in individual to understand how artifacts are used to generate speculation as part of the research frameworks in which they are used, and what needs to be articulated is their dissemination. In HCI speculative artifacts often carry speculative thinking and function as inspiring pieces of design (Filimowicz & Tzankova, 2018). In HCI, artifacts commonly carry speculative thinking, functioning as inspirational pieces. This differs from Dunne and Raby's goal of stimulating discussion. Speculative artifacts in HCI have a longer lifespan. Despite growing interest, few studies systematically examine the introduction of speculative artifacts in HCI.

In responding to this gap, we conducted a scoping literature review of HCI publications that claim to use speculative artifacts. The aim was to provide a comprehensive overview of HCI's current perspectives and approaches to speculative

DOI: 10.1201/9781032693606-11

artifact which dimensions are considered relevant to HCI, and which design and research areas are still under-researched. We first review the key dimensions of artifacts and perspectives on HCI speculative intersections. The results shows different goals, limitations and future directions for integrating speculative artifacts into HCI's multi-epistemological space.

RELATED WORK

INTERSECTIONS BETWEEN HCI AND SPECULATIVE DESIGN

Speculative design has a long history in HCI, reflecting the various ways in which design has been used to consider and shape the future of technology and human interaction. It takes various forms, from proposals to artifacts, from design proposals to artifacts, and as a way to use design to question values and political issues to imagine alternative socio-technical configurations of the world. It enables researchers to move between different forms of "speculation": from predicting the future and forecasting trends for companies to better prepare companies, to critically questioning technology professional's current imagined versions of the future and questioning whether they are correct. Speculative design moves between predicting trends and questioning imagined futures, using rhetorical "fuzziness" to enable critical orientations (Filimowicz & Tzankova, 2018).

Gaver and Martin (2000) coined "speculative design" for conceptual proposals that explore possibilities. Bardzell and Bardzell (2013) examined their ability to both reinforce and resist harmful ideologies, and placed them within the traditions of critical theory from Marx and Nietzsche through the Frankfurt School to an expansion of critical theories in the mid-20th century, including semiotics, poststructuralism, feminism, and psychoanalysis (Bardzell et al., 2012). Pierce et al. (2015) link current speculative design practices within 20th-century avant-garde approaches such as Dada, situationism, and tactical media, as well as activist design approaches. Elsden et al. discuss "Chindogu" (International Chindogu Society, n.d.) which are playful and quirky, designed for entertainment rather than practicality and challenge traditional notions of utility and functionality in design, as a pioneer to speculative design (Elsden et al., 2017). Speculative design was connected to mid-20th-century design and architecture groups, including Archigram and SuperStudio, while discussing Chindogu as a predecessor to speculative design. Beyond art and design, scholars have explored the history of speculative design in areas such as urban planning, which envisions cities, governments, and life in public spaces (DiSalvo et al., 2016).

Wong and Khovanskaya (2018) analyzed the rhetorical work of speculative design methods in promoting the third-wave agendas in HCI and pointed out that speculative design in HCI takes various forms ranging from design proposals to construction artifacts, which are used to imagine alternative sociotechnical configurations of the world as a way to interrogate questions about values and politics through design. Proposed design artifacts as the most common manifestation in critical HCI practice. Speculative design has a multifaceted history in HCI, leveraging diverse forms from proposals to artifacts to imagine and interrogate technological futures, values, and politics across critical theory traditions, avant-garde art, and design approaches.

Multiple Categories of Artifact

Given the diversity of artifacts, HCI researchers advocate elaborating multiple categories for comprehensive understanding. According to Bødker & Klokmose (2012), although previous research on artifact ecology has focused primarily on a static, here-and-now perspective, they explore the dynamics that occur within these ecology when new artifacts are introduced. Material speculation enables the empirical investigation of multiple alternative existences based on the lived experiences of these artifacts (Wakkary et al., 2022). Digital technology has significantly influenced the way people interact and create things, leading to changes in creative activities, as demonstrated by examples such as computer-aided design and rapid prototyping (Jung & Stolterman, 2011). Visitors actively explored the various artifacts and acted as detectives trying to piece together the story of the persona by searching for and examining clues (Leong et al., 2021). From the above-mentioned studies, four categories of artifacts can be generally identified, ranging from the thought and expression level, through the participation and testing level, to the digital and physical context level, and to the assessment reflection level (see Figure 8.1).

Thinking and Expressing: In the early design stages of speculative artifacts, clearly defining the intended user experience is essential. This requires consideration of the impacts on society, culture, ethics, ecology, and politics. The development of conceptual frameworks, predictive models, and analysis tools help to uncover connections and anticipate impacts. For instance, cognitive psychological models can provide insights into the goals of HCI. By elaborating abstract concepts and tools, designers can better anticipate the impacts of artifacts and optimize utility through human-centeredness.

Participation and Test: At the level of participation and testing levels, speculative artifacts often exist long-term to explore research proposals through material speculation. This goes beyond passive creation and actively

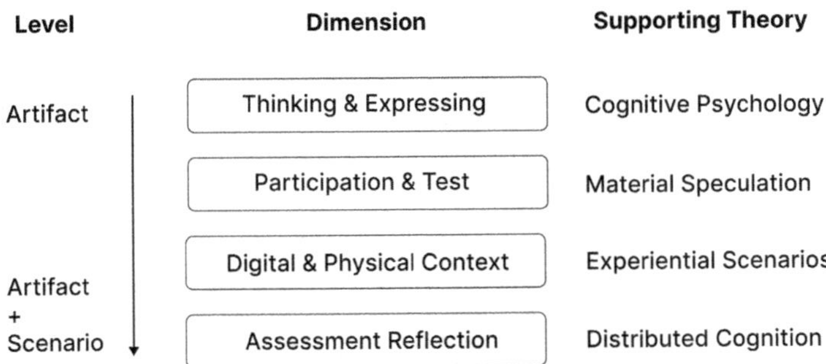

FIGURE 8.1 Four dimensions of application from the artifact level to artifact and scenario.

involves users through participatory and community-based methods. It provides theoretical perspectives, shapes evolving artifacts, constructs experiential knowledge, and enables the exploration of alternative studying alternate hypotheses. In this way, speculative artifacts can reveal new identities, relationships, and processes in socio-technical systems.

Digital and Physical Context: The significance of the digital physical-scene layer in HCI lies in its ability to balance temporal and material constraints, as well as reveal the social and spatial aspects of specific environments. By designing shape-changing artifacts, everyday interactions and experiences can be better supported by creating an artifact that changes shape. In this process, adopting experiential scenarios to design shape-changing artifacts can bridge the gap between the digital and physical realms.

Assessment Reflection: The design principles followed by the Assessment Reflection layer are based on an understanding of the psychological, emotional, and cognitive aspects of technology usage. This process provides guidelines for evaluating the indirect user experience of artifacts. For instance, distributed cognition theory can be used to evaluate aspects that go beyond artifacts. This is about the flow of information across the entire ecosystem of different applications, services, and stakeholders. In such cases, we need to understand and construct user experiences and expectations that go beyond the direct user experience of artifacts and impact other applications, services, and actors in the distributed application ecosystem.

METHOD

A scoping literature review following PRISMA guidelines (Moher et al., 2009) identified HCI publications that adopt speculative artifacts. The ACM Digital Library was searched for "speculative artifact" and "HCI" in titles, abstracts, or keywords and found articles that explicitly claimed relevance. The timeframe was forming the first appearance to February 2023 (Figure 8.2).

This returned 287 initial results. Screening removed unintended matches (n=9), chapter introductions (n=32), inaccessible articles (n=14), and duplicates (n=4), leaving 208 articles. The pool was narrowed down to 184 articles with detailed explanations of speculative artifact adoption in HCI. Articles only mentioning speculative artifacts as an HCI attribute or for future thinking were excluded (n=24).

The 184 articles were classified into either the detailing speculative artifacts/prototypes detailed category (n=124) or categorized into using them loosely (n=60). A link to the full list can be found in the Appendix. Loose uses were included for the quantitative analysis but excluded from the content analysis. The author invited another researcher to read the full text of 124 selected articles separately, then resolved discrepancies through 3 meetings. Open coding of the 124 articles resulted in 24 initial codes. Through discussion, these were consolidated into 12 method and 8 type codes.

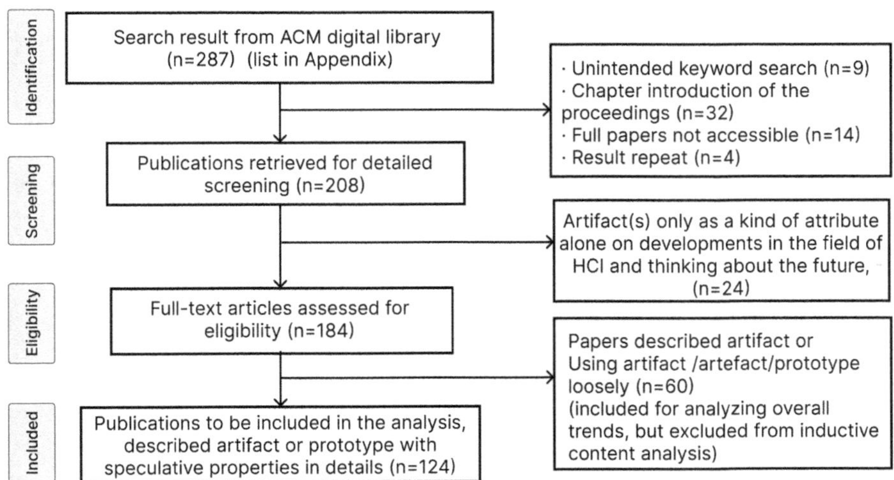

FIGURE 8.2 The publication screening procedure in this study.

To understand trends in HCI research on speculative design, 211 articles from the ACM Digital Library were mapped by year (Figure 8.3). The earliest introduction of the keyword "speculative artifact" in the HCI research field area was in an IBM Watson Research Center conference article published at CHI'89 (Carroll & Kellogg, 1989). Before 2010, speculative artifact articles existed sporadically. Since then, the volumes have significantly increased with notable peaks in 2014, 2016, 2018, 2020, and 2022. The reason for this increase is due to the biennial NordiCHI conference, which brings together researchers and practitioners in the field of HCI from Nordic countries and beyond with a platform to present their latest findings and exchange ideas that can inspire and influence future research directions in this field. A certain number of relevant articles are produced at each conference.

Since the adoption of methods indicates the adoption of practices, we also examined what types of methods that are known as ideal method for generating speculative artifacts have been used in the 124 articles categorized as "described artifact or prototype with speculative properties in details." We found a number of articles in the collection that focus on futuristic tools and use them for analysis. Common methods of conducting experiments and studies mentioned in each article were listed as coding, as detailed in Table 8.1.

Role-play (n=11) was mentioned in 11 articles. New ideas arising from both the challenge of current thinking and possibilities presented by role-playing through new technological the possibilities (e.g., Murray-Rust et al., 2022; Nägele et al., 2018). Role-playing within an interview is a proven method of immersing a participant in a work of fiction and experiencing a current topic from a different perspective (Freytag & Young, 2018). Home investigation approaches were mentioned once in the work of Sabie et al. in 2020, their work was inspired by the work of Lynne Howarth's (Howarth & Knight, 2015). "Indigenous approach" which utilizes surrogates that give voice to

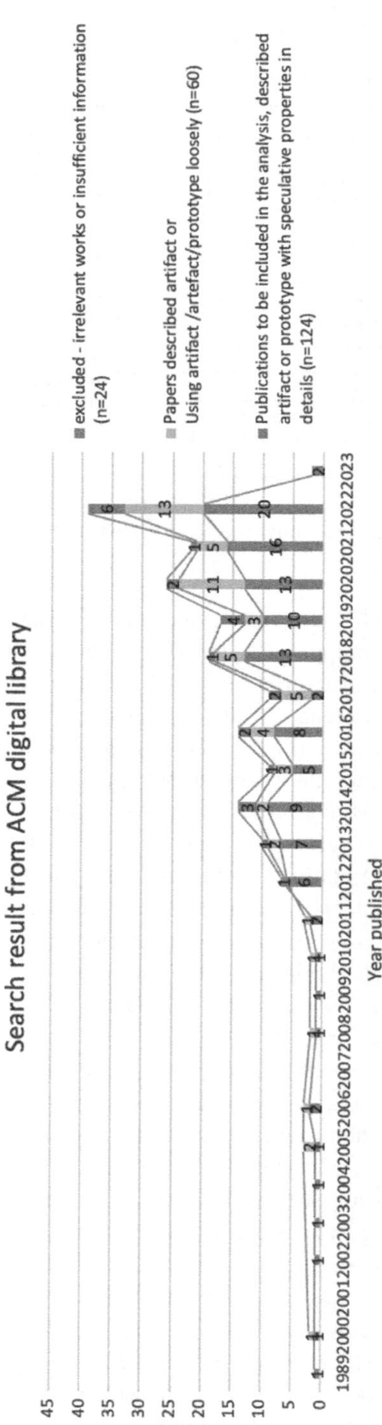

FIGURE 8.3 HCI publications adopting speculative artifacts over time.

TABLE 8.1
The Number of Methods in Our Sample of Literature

Number	Method	Number of articles (%)
1	Role-playing	11 (8.9%*)
2	Investigative approaches for the home	1 (0.8%)
3	Participatory Design	38 (30%)
4	Brainstorming	15 (12%)
5	Social construction of technology (SCOT)	5 (4%)
6	Storytelling	12(9.6%)
7	Heterogeneous discourse	1 (0.8%)
8	Indigenous approach	1 (0.8%)
9	Morphogenetic approach	1 (0.8%)
10	Poetic Interaction Design	2 (1.6%)
11	Futures Cone	2 (1.6%)
12	Technology Probe	1 (0.8%)

* *Percentage is calculated by (no. of articles/124 included articles).*

artifacts and their makers, encompassing references to people, places, relationships, events, techniques, and locations. ***Participatory Design*** (n=38) encourages users to actively be a part of the design (Schuler & Namioka, 1993). Because it is a widely used research method, there are many variations and branches, such as Technology Probe which is an interdisciplinary method for co-designing technologies with users that combines engineering, social science, and design, to understand how users interact with new ones, and these ultimately utilize new technologies (Hutchinson et al., 2003). ***Brainstorming*** (n=15): The attitudes toward the use of brainstorming varied (e.g., Duarte et al., 2018; Ivanov et al., 2022). The pure expression of ideas and the collision of thinking can lead to more inspiration in the early stage of the project, but maintaining objectivity and accompanied by quantitative data support will provide more effective support for the progress of the project.

Social construction of technology (SCOT) (n=5) has emphasized the historical nature of design and that the social, cultural, and institutional context of artifacts plays a key role in shaping the function, intent, and appearance of designs (e.g., Chen et al., 2017; Forlano & Halpern, 2023). **Storytelling** (n=12): Most design futures strive to create a rich, structured immersion in a believable alternative world through the use of multiple media and storytelling techniques, often from the first person (e.g., Kalma et al., 2019). **Heterogeneous discourse** was discussed in an article in the work of Ferri et al. mentioned. They summarized a corpus from which artifacts construct a heterogeneous discourse by contrasting forms and implications from different semantic domains, thus enabling a critical stance (Ferri et al., 2014). The indigenous approach was mentioned in an article (Sabie et al., 2020). They were inspired by Lyne Howarth's work which uses proxies that give voice to artifacts and their makers and include references to people, places, relationships, events, and techniques (Howarth & Knight, 2015). The morphogenetic approach was also used in an article views making as a contingent process or growth to facilitate reflexive and constructive design (Sabie et al., 2020). Poetic interaction design was mentioned in two articles (Kong et al., 2021; Liang & Chang, 2013) which encompasses two features, it provides artifacts for embodied interaction and it evokes poetic imagery. Futures Cone was mentioned in two articles (Forlano & Halpern, 2023; Stead & Coulton, 2022). In Stead and Coulton's work, Futures Cone was used as scaffolding for discussing potential futures starts from an already constrained perspective.

SCOPES OF SPECULATIVE ARTIFACT

In order to make further analysis and comparison of the artifacts in articles, 76 articles were selected, which included pictures of artifacts and detailed descriptions of artifacts. According to the four categories of artifacts (Thinking and Expressing; Participation and Test, Digital and Physical Context; Assessment Reflection) mentioned above, 8 codes were organized under 4 categories: D1 Cognitive pattern: C1 self-speculation, C2 co-speculation; D2 Type of artifact: C3 artifact, C4 artifact with scenario; D3 Function of artifact: C5 test, C6 express; D4 Mode of appearance: C7 digital version, C8 physical version. Each article was coded multiple codes (e.g., Bell et al., 2022, was coded C1, C4, C5, C8). Table 8.2 shows the distribution of articles, but the percentages do not add up to 100% because not all articles belong to a dimension. In the following sections, we explore the dimensions to understand the current integration of speculative thinking and HCI. In the study of cognitive processes related to artifacts, the process by which individuals or groups subjectively reflect on, analyze, and judge a particular problem or topic from different perspectives, leading to self-understanding or shared understanding, can be classified as either individual speculative thinking or collaborative speculative thinking. Individual speculative thinking mainly refers to the thought process, judgment, and reasoning that individuals produce during independent thinking, while collaborative speculative thinking emphasizes group cooperation, joint exploration,

TABLE 8.2
Scopes of Speculative Artifact Adoption

Dimension	Code (C)	Number of articles (%)
D1 Cognitive Pattern	C1 self-speculation	41 (54%*)
	C2 co-speculation	35 (46%)
D2 Type of Artifact	C3 artifact	36 (47%)
	C4 artifact with scenario	40 (53%)
D3 Function of artifact	C5 test	47 (62%)
	C6 express	29 (38%)
D4 Mode of appearance	C7 digital version	27 (36%)
	C8 physical version	49 (64%)

* *Percentage is calculated by (no. of articles/76 included articles).*

and collaborative thinking in order to arrive at more accurate and comprehensive conclusions.

CLUSTERING ANALYSIS

As described in Table 8.2, building upon this foundation, a cluster analysis was conducted to visualize the distribution of the coded artifacts. The Miro online collaborative platform (see Appendix) was utilized to create a visual representation of the speculative artifacts. By taking D1 and D2 as two axles, four quadrants are distinguished, and each quadrant is divided into four regions by D3 and D4. A total of 16 blocks were set up on the Miro board, corresponding to the four dimensions and eight codes established in the coding process. Images of artifacts were extracted from the coded articles and placed in the appropriate blocks based on their assigned categories and with the year of publication. Miro board provides a more intuitive understanding of the distribution of artifact types across the different dimensions, patterns and trends become more apparent (Figure 8.4).

In the self-speculation and artifacts categories, physical and digital artifacts were evenly distributed. Analysis of 13 testing-focused articles found discussions increasingly linking design practice and research in HCI studies. For instance, "research for design" and "research through design" iterative processes enable participants to gain deep understanding of blended physical-digital forms of artifacts (e.g., Belcher et al., 2005; Luu et al., 2018). These examples illustrate the application of speculative techniques to translate sustainable technological futures into actionable design practices.

In the co-speculation and artifacts categories, physical and digital artifacts were again evenly distributed across expressive and verifiable qualities. Analysis of 18 articles revealed that most research was conducted in workshops, courses, etc. using surveys, brainstorming, and other tools. For example, students designed self-ethnographic toolkits for healthier sleep, exploring speculative artifacts for self-exploration (Lockton et al., 2020). In the same category, through a workshop format in 2022, six researchers from IIIT-Delhi explored the importance of intimate relationships as a subset of emotional communication in the HCI field and suggested possible forms of intimate relationships in future artifact designs to explore the relationship between future technologies and intimate relationships, and promote meaningful dialogues in the future. A workshop in 2022 discussed perspectives on intimate relationships in future artifacts and expanded HCI approaches through participatory speculation (Kaur et al., 2023).

In the self-speculation and Artifact+Scenario category, most projects focused on testing physical scenarios, like "Unfabricated" which integrated recyclable materials into design tools to reinvent sustainable smart textiles (Wu & Devendorf, 2020). A small number of self-speculation research projects focus more on expression, such as the Centre for Participatory IT's research project Period Share research project, which represents a typical speculative design. The project proposes a wireless menstrual cup that can automatically quantify and share menstrual data and share it on social networks. Through crowdfunding rhetoric and DIY project form, a specific fictional universe is created, and the potential future of menstrual

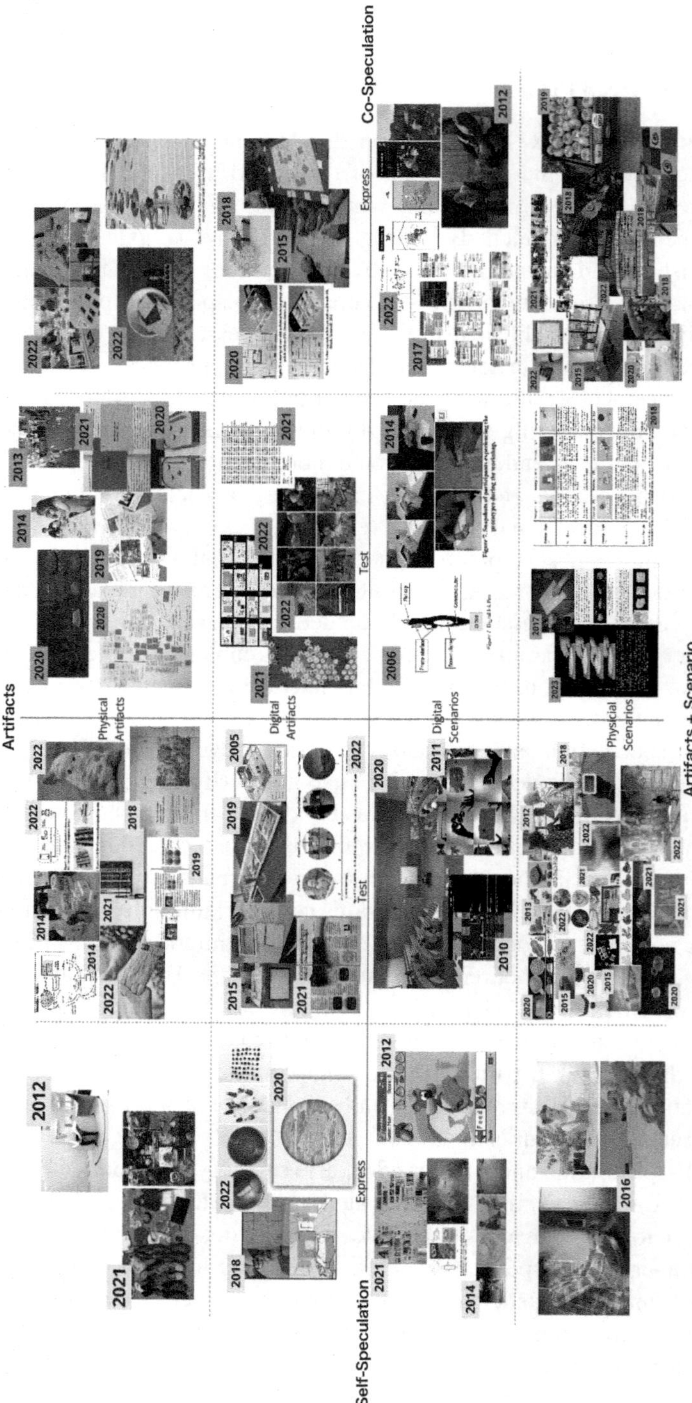

FIGURE 8.4 Quadrant classification of artifacts in this study.

quantification is speculated on. Speculation is encouraged about contemporary technological paradigms such as the politics and culture of self-tracking, sharing, and intimate data around the social, cultural, and political issues of intimate technology (Søndergaard & Hansen, 2016).

In the co-speculation, Artifact + Scenario category, it was found that unlike self-speculation, research projects under co-speculation are more focused on exploring new identities, relationships, and processes through physical scenarios using participatory design and community-based futures methods, thus allowing participants to express their different voices collectively. For example, Liu et al. (2018) conducted, a design research project that explored how collaborative survival shapes interactive artifacts to address ongoing environmental crises. In this context, collaborative survival refers to the concept that human survival is intimately connected to and dependent upon the well-being of numerous other species. To illustrate this idea, the researchers designed wearables for mushroom foraging which reimagined the independent relationship between humans and nature.

DISCUSSION

This study analyzes 124 projects that claim to use speculative design methods in the creation of artifacts, with the aim of understanding how these projects go beyond user-centered approaches to artifacts design. The importance of examining speculative design methods for the HCI field lies in their potential to foster innovative thinking, engage diverse perspectives, and challenge existing paradigms. These findings reveal that artifacts serve multiple roles in these practices aimed at nonexistent futures, effectively connecting reality to future imaginations and providing a dynamic space for oscillation between the two. As shown in Figure 8.4, artifacts differ in their expressive granularity, intended audiences, and the forms of feedback they elicit. Based on these classifications, we can divide these artifacts into three main roles: artifacts as vehicles for co-creation, criticism and extrapolation.

As vehicles for co-creation, researchers recruit groups with diverse backgrounds to participate in the design process. For instance, in the study by Elsden et al. (2017), participants were invited to workshops where they interacted with wearable devices that tracked personal data. This strategy not only introduces external perspectives that differ from internal designers but also encourages richer discussions and the generation of creative ideas. Typically, these engagements manifest as short-term activities, including a series of interviews or workshops that encourage participants to share relevant life experiences and brainstorm collaboratively. Acknowledging that participants may lack creative expressive skills, researchers often use accessible materials such as cardboard, hand drawings, and foam to facilitate simple manipulations.

In their role as vehicles for criticism, artifacts have the task of drawing public attention to specific events or phenomena. By identifying specific issues, researchers use artifacts to provoke discussion, thereby raising awareness and challenging conventional perceptions of everyday products. Hauser et al. (2018) create the "table-non-table", a thought-provoking artifact that challenges traditional notions of utility and domestic objects, serving as a catalyst for reevaluating human-technology relationships and the mediating role of design in everyday life. These artifacts, created through techniques such as 3D printing, laser engraving, and mixed media, are strategically placed in environments where they serve functional purposes. By contrasting sharply with the style of commonly used items, these artifacts blend into everyday environments, utilizing imperfect design to draw the audience's attention to aspects of daily life that have been overlooked.

As vehicles for extrapolation, artifacts empower researchers to envision potential scenarios related to specific issues, grounded in the belief that the presented situations are credible and realistic. Jacobs et al. (2016) illustrate this in "The Prediction Machine" where they present artifacts to simulate future scientific processes and artistic practices. Participants engage with these artifacts, providing feedback that validates the feasibility of underlying assumptions, which is then incorporate into the optimization of proposals. This process requires participants to immerse themselves in the defined context and provide authentic feedback under controlled conditions. Typically, this approach is used to explore long-term product development pathways and propose innovative design projects. The concrete insights and suggestions from participants serve as a critical empirical basis for the design process, ensuring the scenarios presented are relevant and applicable in real-world contexts.

CONCLUSION

This literature review analyses how speculative artifacts are designed and deployed in HCI. By reviewing 124 publications, it can be concluded that the cognitive process of artifacts can be broadly divided into two categories: self-speculation (individual reflection) and co-speculation (group collaboration). Luster analysis revealed multidimensional features that reflect the diverse nature of speculative artifacts. Three main roles of artifacts (co-creation, criticism, and extrapolation) provide a perspective for understanding the interactions between artifacts and their potential for shaping alternative futures. The taxonomy of artifact types, cognitive processes, and roles can influence strategic design and evaluation. However, the analysis is limited by the scoping criteria and coding methodology.

This review contributes to the growing body of work on speculative artifacts and highlights opportunities for targeted research. As HCI grapples with complex socio-technical challenges, researchers and participants can effectively harness the power of artifacts to drive innovation, critical reflection, and meaningful engagement with the visions of the future.

REFERENCES

Aladwan, A., Kelly, R. M., Baker, S., & Velloso, E. (2019). A Tale of Two Perspectives: A Conceptual Framework of User Expectations and Experiences of Instructional Fitness Apps. *Proceedings of the 2019 CHI Conference on Human Factors in Computing Systems*, 1–15. https://doi.org/10.1145/3290605.3300624

Bardzell, J., & Bardzell, S. (2013). What is "critical" about critical design? *Proceedings of the SIGCHI Conference on Human Factors in Computing Systems* (CHI '13, pp. 3297–3306). New York, NY: Association for Computing Machinery. https://doi.org/10.1145/2470654.2466451

Bardzell, S., Bardzell, J., Forlizzi, J., Zimmerman, J., & Antanitis, J. (2012). Critical Design and Critical Theory: The Challenge of Designing for Provocation. *Proceedings of the Designing Interactive Systems Conference* DIS '12, 288–297. New York, NY: Association for Computing Machinery. https://doi.org/10.1145/2317956.2318001

Belcher, J., Aidrus, R., Congleton, B., Hall, D., Hussain, S., Jablonski, M., Klunk, T., & McCrickard, D. S. (2005). NotiFly: Enhancing Design through Claims-Based Personas and Knowledge Reuse. *Proceedings of the 43rd Annual Southeast Regional Conference - Volume 2*, 359–364. https://doi.org/10.1145/1167253.1167339

Bell, F., Al Naimi, L., McQuaid, E., & Alistar, M. (2022). Designing with Alganyl. *Sixteenth International Conference on Tangible, Embedded, and Embodied Interaction*, 1–14. https://doi.org/10.1145/3490149.3501308

Bødker, S., & Klokmose, C. N. (2012). Dynamics in artifact ecologies. *Proceedings of the 7th Nordic Conference on Human-Computer Interaction: Making Sense Through Design*, 448–457. https://doi.org/10.1145/2399016.2399085

Carroll, J. M., & Kellogg, W. A. (1989). Artifact as Theory-Nexus: Hermeneutics Meets Theory-Based Design. *CHI'89: Proceedings of the SIGCHI Conference on Human Factors in Computing Systems*, 7–14. https://doi.org/10.1145/67450.67452

Chen, W., Crandall, D. J., & Su, N. M. (2017). Understanding the Aesthetic Evolution of Websites: Towards a Notion of Design Periods. *Proceedings of the 2017 CHI Conference on Human Factors in Computing Systems*, 5976–5987. https://doi.org/10.1145/3025453.3025607

DiSalvo, C., Jenkins, T., & Lodato, T. (2016). Designing Speculative Civics. *Proceedings of the 2016 CHI Conference on Human Factors in Computing Systems*, 4979–4990. https://doi.org/10.1145/2858036.2858505

Dourish, P. (2004). What we talk about when we talk about context. *Personal and Ubiquitous Computing*, *8*(1), 19–30. https://doi.org/10.1007/s00779-003-0253-8

Duarte, E. F., Gonçalves, F. M., & Baranauskas, M. C. C. (2018). InstInt: Enacting a Small-Scale Interactive Installation through Co-design. *Proceedings of the 30th Australian Conference on Computer-Human Interaction*, 338–348. https://doi.org/10.1145/3292147.3292158

Dunne, A. (1999). *Hertzian tales: Electronic products, aesthetic experience, and critical design*. MIT Press.

Dunne, A., & Raby, F. (2013). *Speculative everything: Design, fiction, and social dreaming*. The MIT Press.

Elsden, C., Chatting, D., Durrant, A. C., Garbett, A., Nissen, B., Vines, J., & Kirk, D. S. (2017). On Speculative Enactments. *Proceedings of the 2017 CHI Conference on Human Factors in Computing Systems*, 5386–5399. https://doi.org/10.1145/3025453.3025503

Ferri, G., Bardzell, J., Bardzell, S., & Louraine, S. (2014). Analyzing critical designs: Categories, Distinctions, and Canons of Exemplars. *Proceedings of the 2014 Conference on Designing Interactive Systems*, 355–364. https://doi.org/10.1145/2598510.2598588

Filimowicz, M., & Tzankova, V. (Eds.). (2018). *New directions in third wave human-computer interaction: Volume 2 - methodologies*. Springer International Publishing. https://doi.org/10.1007/978-3-319-73374-6

Forlano, L. E., & Halpern, M. K. (2023). Speculative Histories, Just Futures. *ACM Trans. Comput.-Hum. Interact.* https://doi.org/10.1145/3577212

Freytag, P. V., & Young, L. (Eds.). (2018). *Collaborative research design*. Springer Singapore. https://doi.org/10.1007/978-981-10-5008-4

Gaver, B., & Martin, H. (2000). Alternatives: Exploring information appliances through conceptual design proposals. *In Proceedings of the SIGCHI Conference on Human Factors in Computing Systems (CHI '00)*, 209–216. Association for Computing Machinery. https://doi.org/10.1145/332040.332433

Harrison, S., Tatar, D., & Sengers, P. (2007). The Three Paradigms of HCI. *Alt. Chi. theIG-CHI Conference on Human Factors in Computing Systems San Jose, California,*

Hauser, S., Wakkary, R., Odom, W., Verbeek, P. P., Desjardins, A., Lin, H., Dalton, M., Schilling, M., & de Boer, G. (2018). Deployments of the table-non-table: A Reflection on the Relation Between Theory and Things in the Practice of Design Research. *In Proceedings of the 2018 CHI Conference on Human Factors in Computing Systems,*1–13. https://doi.org/10.1145/3173574.3173775

Howarth, L. C., & Knight, E. (2015). To every artifact its voice: Creating surrogates for hand-crafted indigenous objects. *Cataloging & Classification Quarterly, 53*(5–6), 580–595. https://doi.org/10.1080/01639374.2015.1008719

Hutchinson, H., Mackay, W., Westerlund, B., Bederson, B. B., Druin, A., Plaisant, C., Beaudouin-Lafon, M., Conversy, S., Evans, H., Hansen, H., Roussel, N., Eiderbäck, B., Lindquist, S., & Sundblad, Y. (2003). Technology probes: Inspiring design for and with families. In *Proceedings of the SIGCHI Conference on Human Factors in Computing Systems*, 17–24. https://doi.org/10.1145/642611.642616

International Chindogu Society.(n.d.). Chindogu: Official home of the International Chindogu Society. Retrieved April 14, 2023, from http://chindogu.com/ics/

Ivanov, A., Au Yeung, T., Blair, K., Danyluk, K., Freeman, G., Friedel, M., Hull, C., Hung, M. Y.-S., Pratte, S., & Willett, W. (2022). One Week in the Future: Previs Design Futuring for HCI Research. *CHI Conference on Human Factors in Computing Systems*, 1–15. https://doi.org/10.1145/3491102.3517584

Jacobs, R., Benford, S., & Luger, E. (2016). The Prediction Machine: Performing Scientific and Artistic Process. *In Proceedings of the 2016 ACM Conference on Designing Interactive Systems*, 497–508. https://doi.org/10.1145/2901790.2901825

Jung, H., & Stolterman, E. (2011). Form and materiality in interaction design: A new approach to HCI. *In CHI '11 Extended Abstracts on Human Factors in Computing Systems*, 399–408. Association for Computing Machinery. https://doi.org/10.1145/1979742.1979619

Kalma, A., Ploderer, B., Sitbon, L., & Brereton, M. (2019). Probing Yarns about Ageing and Making. *Proceedings of the 31st Australian Conference on Human-Computer-Interaction*, 173–183. https://doi.org/10.1145/3369457.3369472

Kaur, J., Devgon, R., Goel, S., Singh, A., Monteiro, K., & Singh, A. (2023). Future of Intimate Artefacts: A Speculative Design Investigation. *Proceedings of the 13th Indian Conference on Human-Computer Interaction*, 57–66. https://doi.org/10.1145/3570211.3570216

Kong, B., Liang, R. -H., Liu, M., Chang, S.-H., Tseng, H. -C., & Ju, C. -H. (2021). Neuromancer Workshop: Towards Designing Experiential Entanglement with Science Fiction. *Proceedings of the 2021 CHI Conference on Human Factors in Computing Systems*, 1–17. https://doi.org/10.1145/3411764.3445273

Leong, T. W., Su, C.-S., Liang, R.-H., & Tsai, W.-C. (2021). Experiential Persona: Towards supporting richer and unfinalized representations of people. *Extended Abstracts of the 2021 CHI Conference on Human Factors in Computing Systems*, Article 222, 1–6. https://doi.org/10.1145/3411763.3451700

Liang, R. -H., & Chang, H. -M. (2013). Hypnotist Framing: Hypnotic Practice as a Resource for Poetic Interaction Design. *Proceedings of the 6th International Conference on Designing Pleasurable Products and Interfaces*, 241–250. https://doi.org/10.1145/2513506.2513532

Liu, J., Byrne, D., & Devendorf, L. (2018). Design for Collaborative Survival: An Inquiry into Human-Fungi Relationships. *Proceedings of the 2018 CHI Conference on Human Factors in Computing Systems*, 1–13. https://doi.org/10.1145/3173574.3173614

Lockton, D., Zea-Wolfson, T., Chou, J., Song, Y. (Antonio), Ryan, E., & Walsh, C. (2020). Sleep Ecologies: Tools for Snoozy Autoethnography. *Proceedings of the 2020 ACM Designing Interactive Systems Conference*, 1579–1591. https://doi.org/10.1145/3357236.3395482

Luu, T., van den Broeck, M., & Søndergaard, M. L. J. (2018). Data Economy: Interweaving Storytelling and World Building in Design Fiction. *Proceedings of the 10th Nordic Conference on Human-Computer Interaction*, 771–786. https://doi.org/10.1145/3240167.3240270

Malpass, M. (2016). Critical design practice: Theoretical perspectives and methods of engagement. *The Design Journal*, *19*(3), 473–489. https://doi.org/10.1080/14606925.2016.1161943

Mankoff, J., Rode, J. A., & Faste, H. (2013). Looking past yesterday's tomorrow: Using futures studies methods to extend the research horizon. *Proceedings of the SIGCHI Conference on Human Factors in Computing Systems* (CHI '13, pp. 1629–1638). Association for Computing Machinery, New York, NY, USA. https://doi.org/10.1145/2470654.2466216

Moher, D. G., Liberati, A., Tetzlaf, J., & Altman. (2009). Preferred reporting items for systematic reviews and meta-analyses: the PRISMA statement. *PLoS Med*, 6(7), e1000097.https://doi.org/10.1371/journal.pmed.1000097

Murray-Rust, D., Elsden, C., Nissen, B., Tallyn, E., Pschetz, L., & Speed, C. (2022). Blockchain and beyond: Understanding blockchains through prototypes and public engagement. *ACM Transactions on Computer-Human Interaction*, *29*(5), 1–73. https://doi.org/10.1145/3503462

Nägele, L. V., Ryöppy, M., & Wilde, D. (2018). PDFi: Participatory Design Fiction with Vulnerable Users. *Proceedings of the 10th Nordic Conference on Human-Computer Interaction*, 819–831. https://doi.org/10.1145/3240167.3240272

Pierce, J., Sengers, P., Hirsch, T., Jenkins, T., Gaver, W., & DiSalvo, C. (2015). Expanding and Refining Design and Criticality in HCI. *Proceedings of the 33rd Annual ACM Conference on Human Factors in Computing Systems*, 2083–2092. https://doi.org/10.1145/2702123.2702438

Sabie, D., Sabie, S., & Ahmed, S. I. (2020). Memory through Design: Supporting Cultural Identity for Immigrants through a article-Based Home Drafting Tool. *Proceedings of the 2020 CHI Conference on Human Factors in Computing Systems*, 1–16. https://doi.org/10.1145/3313831.3376636

Schuler, D., & Namioka, A. (Eds.). (1993). *Participatory design: Principles and practices*. CRC Press. https://doi.org/10.1201/9780203744338

Smith, R. C., Iversen, O. S., Hjermitslev, T., & Lynggaard, A. B. (2013). Towards an Ecological Inquiry in Child-Computer Interaction. *Proceedings of the 12th International Conference on Interaction Design and Children*, 183–192. https://doi.org/10.1145/2485760.2485780

Søndergaard, M. L. J., & Hansen, L. K. (2016). PeriodShare: A Bloody Design Fiction. *Proceedings of the 9th Nordic Conference on Human-Computer Interaction*, 1–6. https://doi.org/10.1145/2971485.2996748

Stead, M., & Coulton, P. (2022). Sustainable Technological Futures: Moving beyond a One-World-World perspective. *Nordic Human-Computer Interaction Conference*, 1–17. https://doi.org/10.1145/3546155.3547283

Sterling, B. (2005). *Shaping things*. Cambridge, MA: MIT Press.

Tharp, B. (2013). Discursive design basics: Mode and audience. Nordes 2013: Experiments in Design Research. https://doi.org/10.21606/nordes.2013.051

Wakkary, R., Odom, W., Hauser, S., Hertz, G., & Lin, H. (2016). A short guide to material speculation: Actual artifacts for critical inquiry. *Interactions*, *23*(2), 44–48. http://dx.doi.org/10.1145/2889278

Wakkary, R., Oogjes, D., & Behzad, A. (2022). Two Years or More of Co-speculation: Polylogues of Philosophers, Designers, and a Tilting Bowl. *ACM Transactions on Computer-Human Interaction*, 29(5), Article 47, 1–44. https://doi.org/10.1145/3514235

Wong, R. Y., & Khovanskaya, V. (2018). *Speculative design in HCI: From corporate imaginations to critical orientations.* Springer International Publishing, 175–202. https://doi.org/10.1007/978-3-319-73374-6_10

Wu, S., & Devendorf, L. (2020). Unfabricate: Designing Smart Textiles for Disassembly. *Proceedings of the 2020 CHI Conference on Human Factors in Computing Systems*, 1–14. https://doi.org/10.1145/3313831.3376227

APPENDIX

The complete list of all 287 publications and their categorization in this study can be found at https://rich-seeder-dae.notion.site/List-of-articles-used-in-the-scoping-review-616e206874544ad98d3ba1ea672fc935

Miro Quadrant classification can be found at https://miro.com/welcomeonboard/ckNtVHNuSDBmZXJKVkluR2VNeUFaSlJDbjFzSjc0UW1sbDFSN2pDUUFIcj-ViSEZyRXJJSWpQY2ptWWxZVWdmeHwzNDU4NzY0NTIzNzY3ODcyNzc5fDI=?share_link_id=178613485664

9 Interactive Narrative and Scenario Building

Jiawei Li and Yin Li

THE RELATIONSHIP BETWEEN INTERACTIVE NARRATIVES AND SCENARIO BUILDING

Definition of Interactive Narratives and Scenario Construction

As a narrative form, interactive narratives emphasize the interactivity between the audience and the story, in contrast to traditional linear narratives (Murray, 1997). In interactive narratives, the audience is no longer a passive recipient of information but becomes an active participant in the story, capable of influencing the development and outcome of the plot. This form of storytelling breaks away from the one-dimensionality of traditional narratives, allowing the audience to make choices, decisions, or interactions within the story, thus personalizing their narrative experience. Through multiple storylines, branching structures, or decision nodes, interactive narratives provide audiences with diverse choices, enabling the same story to unfold in different plot directions. In familiar practical cases, interactive narratives have been widely applied in areas such as digital games, virtual reality (VR), film, and online storytelling. "Black Mirror: Bandersnatch" is a typical example that shapes different storylines based on audience choices, showcasing the potential of interactive narratives. Additionally, the game "Resident Evil: Village" constructs a dynamic story world through player choices and actions.

However, interactive narratives alone are not sufficient to create a profound storytelling experience. The background of the story, serving as the stage for the unfolding plot, plays an undeniable role in narrative experience. This leads us to the discussion of scenario construction. Scenario construction plays a crucial role in storytelling, as it can create specific scenes by combining elements such as environment, characters, and time, providing a more vivid and concrete backdrop for the narrative (Ryan, 2006). Scenario construction is not only the background setting required for storytelling but also a means to enhance the audience's perception of the narrative experience through scene creation. In scenario construction, attention should be given not only to textual narration but also to various sensory stimuli such as visual and auditory elements. Detailed scenario descriptions can immerse the audience more deeply into the story, allowing them to feel the emotions, atmosphere, and conflicts within.

Interactive narratives and scenario construction complement each other, collectively building a more dimensional and interactive storytelling experience. Interactive narratives, through audience participation, transform scenario construction from a static background into a dynamic "stage" that evolves with audience choices and

DOI: 10.1201/9781032693606-12

decisions. Simultaneously, scenario construction provides interactive narratives with more detailed and realistic backgrounds, enabling audiences to immerse themselves better in the story through interaction.

The construction of narratives is often intertwined with scenario construction, as Zurlo considers it a form of knowledge or world representation that can be transformed into perception, understanding, and action (Zurlo & Cautela, 2014). In the field of future research, scenario construction specifically refers to the core conceptualization phase in scenario planning processes, serving as a crucial method and technology for building diverse worlds and alternative futures. Designers often prefer to emphasize envisioning user-centered solutions through the use of scenario tools. For designers, a "scenario" is typically understood as a story of how an individual interacts with a product or service in their daily life. This design perspective relies on experiential or fictional details, focusing on interactions between people and objects—what is currently happening and what might happen in the future. Scenario techniques commonly used in design practice include creating storyboards and videos that narrate someone's future experiences and highlight key touchpoints when engaging with an organization's product or service. By fostering interaction among diverse stakeholders through more narrative and visual scenarios, this approach drives rapid iteration of design prototypes and solutions.

The Roles and Importance of Interactive Narration and Scenario Construction in HCI

Interactive narration and scenario construction are two crucial research directions in the field of human-computer interaction (HCI). They play a key role in enhancing user experience, increasing user engagement, and driving the development of HCI research. In HCI, we need to consider not only textual narration but also various sensory stimuli such as visual and auditory elements. Through detailed scenario descriptions, users can immerse themselves more deeply in the story, experiencing emotions, atmosphere, and conflicts. The complementary features of interactive narration and scenario construction collectively build a more vivid and interactive narrative experience.

As mentioned in the previous chapter, interactive narration primarily focuses on guiding user interaction with computer systems through the form of stories. Leveraging humans' innate preference for stories, it presents complex tasks or information in a narrative format, thereby reducing user learning costs and improving efficiency and satisfaction in task completion (Green & Bavelier, 2012). Scenario construction, on the other hand, is a design method that helps designers and users understand and predict the behavior of systems by creating and using contexts or situations (Wixon & Hollands, 2005). In HCI, the importance of interactive narration and scenario construction is mainly reflected in the following aspects:

First, they can enhance user engagement and satisfaction. Interactive narration, through storytelling, captures user attention and arouses interest, thereby increasing user engagement. Simultaneously, scenario construction helps users better understand and predict system behavior, thereby enhancing satisfaction (Hassenzahl &

Tractinsky, 2006). Second, they can reduce user learning costs. Interactive narration simplifies complex tasks or information through storytelling, making it easier for users to understand and grasp. Scenario construction, by creating and using contexts or situations, helps users predict system behavior, reducing their learning burden (Nielsen, 1993). Moreover, to some extent, they drive the development of HCI research. Interactive narration and scenario construction provide new perspectives and methods for HCI research, enabling researchers to understand and address interaction problems between users and computer systems from different angles (Gaver & Schwartz, 2003).

Carroll introduced the concept of scenarios into the HCI field in the mid-1990s, establishing a series of scenario-based design methods and techniques (Carroll, 1995). Unlike the use of scenarios in strategic foresight, which replaces the future, scenario-based design emphasizes specific descriptions of future system use through narration in the early stages of the development process.

The introduction of the scenario concept shifted the focus of HCI design work from defining system operations (i.e., functional specifications) to describing how people use systems to accomplish work tasks and other activities. Some scholars narrowly define scenarios as stories, emphasizing that in HCI, scenarios consist of an environment or situational state, one or more participants with personal motivations, knowledge, and abilities, and a series of tools and artifacts that participants perceive and manipulate. In HCI, scenarios describe a series of actions and events that lead to the outcome of HCI in a certain contextual context.

Exploring the Prospects of Interactive Narration and Scenario Construction in the Digital Futures

In this digital era, interactive narration and scenario construction present endless possibilities, painting a creative picture of the future. With the increasing maturity of artificial intelligence (AI) technology, designers will be able to leverage intelligent algorithms to analyze user preferences and behavior more accurately, injecting more personalized and captivating elements into scenarios. As Herbert A. Simon stated in his classic work "The Sciences of the Artificial," "Everyone designs who devises courses of action aimed at changing existing situations into preferred ones", emphasizing that design is not merely a problem-solving process but also a process of creating solutions to transform current situations into desired outcomes (Simon, 1969). The rise of AI undoubtedly provides designers with more powerful innovative tools, propelling both fields into entirely new creative territories.

On the other hand, the emergence of augmented reality (AR) and VR technologies will bring unprecedented immersion to both interactive narration and scenario construction. The application of AR technology allows virtual elements to seamlessly integrate into the user's real environment, creating a more realistic and captivating visual experience. The boundaries between the digital and the real become blurred, and users feel as if they have entered a world full of imagination. The use of blockchain technology provides new possibilities for credibility and copyright protection. Its decentralized and tamper-proof characteristics offer designers effective means to ensure the authenticity and originality of digital scenarios. This not only provides a more reliable creative environment for creators but also brings a more secure digital experience for users.

However, these transformations go beyond the technological realm. The rise of social media makes digital storytelling more diverse and multidirectional. Users can share their interactive experiences on social platforms, building a creative and resonant community. Designers need to harness the power of this trend, using social media to stimulate user participation and break free from the constraints of traditional narratives.

The ubiquity of digitization also comes with a series of challenges. The existence of the digital divide requires us to consider how to make digital creativity more widespread and how to involve more people in this digital feast. Enhancing digital literacy, addressing digital inequality, and recognizing cultural differences will be among the topics designers need to ponder. Designers need narrative skills to quickly construct such scenarios, not only through traditional storytelling tools in the design process but also to make the possibilities of imagination and conception more vivid and concrete. More interactive ways of scenario narration within the realm of digital construction have advantages in immersive collaborative innovation. On the one hand, it facilitates rapid internal dialogue within design teams through scenario building, and on the other hand, it provides users with a stronger foundation for multimodal experiences. Driven by the evaluation of digital future usability, designers need to consider how to organically combine text, images, audio, video, and virtual elements to create a more rich and captivating narrative experience.

BASIS OF INTERACTIVE NARRATIVE

THE CONCEPT OF INTERACTIVE NARRATION

Interactive narration is an intriguing narrative form that has emerged in the digital age, characterized by empowering the audience with greater participation, making them co-creators of the story through interactivity. This section delves into the concept of interactive narration, covering its definition, features, and its disruptive impact on traditional narrative modes.

At the core of interactive narration is the interactive relationship between the audience and the story. Janet Murray (1997), in her work "Hamlet on the Holodeck," mentions that interactive narration is a dynamic and non-linear narrative form, enabling the audience to participate in the development of the story through choices, exploration, and interaction. This definition highlights interactivity as a key element of interactive narration, transforming the audience from passive observers into active participants. Its non-linear structure, diverse narrative paths, real-time feedback, and personalized experiences are particularly prominent. Murray's research emphasizes a "focus on the participant," highlighting the active role of the audience in storytelling. This feature makes interactive narration a more dynamic and flexible narrative form, contrasting sharply with traditional linear narratives (Murray, 1997).

The rise of interactive narration is not only a technological innovation but also a profound challenge to traditional narrative paradigms. Espen Aarseth (1997), in "Cybertext: Perspectives on Ergodic Literature," emphasizes that the non-linear nature of interactive narration disrupts the inherent structure of traditional narratives, compelling creators to rethink the forms and methods of storytelling. This disruption extends beyond technological implementation to touch upon the audience's sense of participation and agency in storytelling (Aarseth, 1997).

Illustrating the Structure of Interactive Narration through Real Cases

With the development of the gaming industry, the structure of interactive narration has gradually become a crucial element in game design. As mentioned earlier, unlike traditional linear narratives, interactive narration emphasizes player participation and choices, allowing the story to evolve based on player decisions. This section analyzes in-depth the characteristics and advantages of the interactive narrative structure through real case studies. In recent years, many well-known games have incorporated interactive narrative structures, with the "Cardboard House" series standing out as an exemplary case. This game allows players to accumulate political capital through various means, influencing the course of the storyline. For instance, in a particular level, players can choose to ally with a political ally or adopt other strategies, directly impacting the subsequent game plot and ending. This narrative approach transforms players from passive story recipients into crucial participants in the story's development. Additionally, a game named "Choose Your Own Adventure" also utilizes the techniques of interactive narration. In this game, players need to make choices based on the progression of the story, and these choices determine the direction and conclusion of the story.

Interactive narration not only demonstrates its notable features in the gaming industry but also actively participates in the entertainment industry, particularly in film and television. A prominent example is the interactive novel "Black Mirror: Bandersnatch." This work is part of the "Black Mirror" series, focusing on themes of science fiction and technology, and was released on Netflix in 2018. In this interactive movie, viewers need to make choices at crucial moments, directly influencing the development and outcome of the story. This interactivity not only provides diverse plot directions but also prompts viewers to contemplate moral and ethical choices.

From a psychological perspective, the structure of interactive narration offers viewers a more personalized and engaging experience. According to psychologist Csikszentmihalyi's theory of "flow," people feel pleasure and fulfillment when engaged in highly interactive and challenging tasks. Interactive narration, by allowing viewers to make choices in the story, creates a narrative experience that aligns more closely with the "flow" state, making viewers more invested and immersed. Furthermore, the structure of interactive narration involves the non-linearity of storytelling, breaking free from the constraints of traditional linear narratives. Compared to traditional linear narrative structures, interactive narration structures have significant advantages. First, they enhance the playability of games. Due to the diversification of storylines, players can experience different plots when replaying the game, thus extending the game's lifespan. Second, interactive narration structures increase the immersion of games. Player choices significantly impact the story's development, making it easier for players to immerse themselves in the game world. Finally, this narrative approach is conducive to shaping more realistic and complex character images. Because player decisions affect character relationships and plot development, the characters in the game become more three-dimensional and multifaceted.

However, the structure of interactive narration also faces some challenges. Research indicates that viewer choices may lead to the dispersion and confusion of the story, making it challenging to maintain consistency and depth in the narrative. Additionally, viewers' choices may be influenced by subjective preferences and emotions rather than objective considerations, leading to deviations in the story's development. This issue needs careful attention and resolution when designing interactive narratives.

In summary, the structure of interactive narration has a wide range of applications in the entertainment industry. Through the analysis of real case studies, we can see the advantages of this narrative approach: increased playability, enhanced immersion, and the shaping of more realistic character images. With continuous technological advancements, we look forward to the emergence of more excellent games with interactive narrative structures, providing players with richer and more diverse gaming experiences.

SCENARIO CONSTRUCTION FRAMEWORK

THE ROLE AND SIGNIFICANCE OF SCENARIO CONSTRUCTION IN HCI

In the current context of HCI emphasizing post-humanism and the mapping of social values, there is a potential for in-depth exploration of scenario construction under the terms of worldview mapping and world building. Traditional HCI is more often combined with predictive scenario construction, and visible futures in future research are also largely based on such baseline scenarios. By guiding and discussing the concept of world building pluralism, the idea of world building is conceptualized as a creative act involving all human capabilities, leading to more imaginative scenario development (Vervoort et al., 2015). In science fiction and fantasy stories, world building is a commonly used narrative technique. It involves creating a detailed and immersive fictional world with its own unique rules, culture, and history. Although world building is often closely connected with storytelling, it can also serve as a powerful tool for designing the future, providing a framework for envisioning and constructing plausible future scenarios.

In 2015, Peter von Stackelberg and Alex McDowell proposed that world building is the process of constructing a complete and rational imaginary world, serving as the background for a story. It is "the creation of an imagined world with coherent geography, society, culture, and other features" (Stackelberg & Alex, 2015). World building provides designers with a detailed set of background rules, develops a larger reality beyond a single story, and potentially offers a deeper understanding of the underlying systems that drive these worlds. World building can provide information for foresight practices and help establish stronger relationships between foresight and system design. Leah Zaidi likens world building to the storytelling of social constructivism and system design (Zaidi, 2017). Similar to how our social ecosystem is interrelated, co-evolving, and "interconnected in an endless adaptive cycle of growth, accumulation, and 'recomposition'," science fiction worlds instill a sense of completeness. World building does not need to be narrow or centered around artifacts; it can and should aspire to be systematic. Although world building tends to be systematic and has connections with foresight in the context of science fiction, it is an underexplored intentional design practice with potential applications in the real world. Raven and Elahi point out, "There is almost no literature that applies the narrative strategies and logic understood by writers, filmmakers, and cultural scholars to the methods used by future scholars and practitioners in creating their final outputs" (Raven & Elahi, 2015). Therefore, the scenario construction framework can guide designers to think strategically not only about task-based scenarios, prototypes, user interface metaphors, usability standards, and user needs but also about envisioning system visions and design principles. By considering the complexity of systems through scenario thinking, designers can explore how the designed HCI interface can connect humans with society, environment, and ecology. Beyond system design, this approach propels the meaning construction of a digital future, better mapping the values of designers and their project stakeholders.

The Relationship Between Scenario Construction and User Experience

The relationship between scenario construction and user experience profoundly influences the design of digital products and user engagement. Serving as the cornerstone of digital prototype design, scenario construction provides designers with a framework to understand user needs and the environment. This process encompasses user behaviors, needs, expectations, and interactions with the system in specific contexts, laying a solid foundation for user experience design.

During scenario construction, designers must consider the actual scenarios in which users will be using the product or system. This involves various aspects such as the user's environment, emotional state, and task objectives. Through meticulous scenario construction, designers can better simulate and understand user behavior in specific contexts, offering robust guidance for user experience design. For instance, anticipating scenarios where users operate the product in high-pressure work environments allows designers to provide targeted, concise, intuitive, and efficient interfaces, ensuring smooth product use when dealing with work-related stress. Scenario construction not only plays a crucial role in the early stages of design but can also be extensively applied during user testing. By simulating real scenarios, designers can more accurately observe user reactions and interactions in specific usage contexts. This aids in identifying potential issues and improving design, enhancing product usability and user satisfaction. For example, simulating the shopping experience in user testing allows designers to better understand potential problems users may encounter during the shopping process, leading to corresponding optimizations and improvements.

In the design of Google Maps, scenario construction became a key element in optimizing the user experience. The design team conducted in-depth research into user needs for map applications in different usage scenarios, such as walking navigation, driving route planning, and urban exploration, formulating detailed scenario constructions. For instance, in the walking navigation scenario, considering that users may find it challenging to view the phone screen while walking, the design team implemented features like voice navigation and vibration prompts to offer a more intuitive and safe navigation experience.

The user experience design team then based their work on these constructed scenarios. They optimized the interface of the map application to ensure it provides intuitive and user-friendly interaction methods in different usage scenarios. For example, during driving navigation, the design team improved the convenience of using the map in the vehicle by adding large and clear turn prompts, as well as displaying real-time traffic information. This scenario-based user experience design allows Google Maps to better adapt to users' daily lives, providing services that align more closely with actual needs.

Furthermore, normative scenario narrative documents play a crucial role in this process. These documents provide detailed descriptions of specific usage scenarios, helping designers better understand user behavior patterns and expectations. These documents also serve as reference points for iterative feedback and evaluation. Designers can continually adjust and enhance scenario construction to adapt to user feedback and evolving needs. User experience design, based on scenario construction, aims to create products or systems that are satisfying, easy to use, and enjoyable for users. By deeply understanding user needs and behavior patterns, designers can optimize the interaction design, interface layout, and feature settings of the product to offer an experience that aligns with user expectations. For example, through scenario construction focused on users' usage of a specific application on mobile devices, designers can better optimize the mobile user interface of the application, ensuring users enjoy a smooth and consistent experience in different contexts.

In the design process of digital products, scenario construction and user experience design form a mutually dependent and reinforcing relationship. Scenario construction provides the foundation for user experience design, while user experience design enriches the details of scenario construction through continuous optimization and refinement. This collaborative interaction allows designers to comprehensively consider user needs, polishing digital products that closely align with real-world usage scenarios. By delving into the relationship between scenario construction and user experience, designers can better guide product development, creating a more outstanding user experience.

FUTURE TRENDS OF INTERACTIVE NARRATIVES AND SCENARIO CONSTRUCTION

Against the backdrop of metamodernism and post-humanism in the development of user experience, the design community is compelled to contemplate the enormous uncertainty and rapid changes brought about by information technology, globalization, climate change, and biotechnology to humans and related socio-technical systems. Designers and their affiliated groups face significant challenges in mapping and correlating networks according to actor-network theory (ANT). The effectiveness of tools and technologies for diverse and alternative scenarios in guiding the design and development of new systems, and whether they can drive designers to recognize that narrative is the cornerstone of human perception, actively applying systems thinking to embrace the complexity of HCI, and daring to explore factors beyond systems and their environments, represents one of the significant challenges in combining narrative concepts with design.

In 1995, American scholar Nicholas Negroponte, in his book "Being Digital," proposed, "Humans are about to live in a virtual, digitized living space, where people will use digital technology for information dissemination, communication, learning, work, and other activities. This is the digitized world" (Negroponte, 1995). Looking at the attributes of digital life, it can be characterized by four features: decentralization of power, globalization, the pursuit of harmony, and empowerment. Based on this, an increasing number of network users can wield the power of tools to influence and change their digital lives. The tools and technologies for designing fictions proposed by Bruce Sterling in 2005 are a significant manifestation of scenario construction in the field of design. With the emergence of Brian David Johnson's science fiction prototype method and the popularization and promotion of design fiction by Bleecker, narrative scenarios have greatly contributed to the development and application of HCI. For futurists and practitioners in the foresight field, storytelling is considered the best way to stimulate imagination and possible futures (Ramirez et al., 2010). Drawing from the usage methods of scenarios in futures studies and focusing on how the form of stories changes with the rise of cross-media and new tools, a "new narrative" is proposed to drive the deep integration of interactive narratives and scenario construction under the semantic design.

With the intervention of new ICT technologies, urban environments have become more sensory, interactive, and flexible. The integration of physical space and intangible networks in the city supports human activities at various levels, forming a highly complex ecosystem. Humans increasingly desire to seek a balance of technology in complex systems from a new perspective, hoping to find a subtle and authentic "digital well-being." Take the emerging field of the metaverse as an example. The metaverse anticipates a parallel internet space running alongside the real world, implying that the VR network world has become a new social ecosystem. Based on the changes in life and communication in the metaverse, from the virtual dimension to the real dimension, helping users rebuild a virtual symbiotic environment suitable for the fusion of humanities and technology in a world where carbon- and silicon-based civilizations coexist becomes a new challenge in the design field.

Future scenario construction is an essential tool in the innovative design process for informing, validating, and gaining approval for design decisions. With the help of interactive narratives and scenario construction technology, designers can not only focus on product-service systems but also engage in effective dialogues between future visions and strategic formulation for organizations. Designers use interactive narratives to dispel skepticism among project stakeholders about internal and external change factors. Through the scenarios and forward-looking content derived from narratives, they collectively explore a mutually agreed-upon future.

REFERENCES

Murray, J. H. (1997). Hamlet on the Holodeck: The Future of Narrative in Cyberspace. The Free Press.

Ryan, M. L. (2006). Avatars of Story. University of Minnesota Press.

Zurlo, F., & Cautela, C. (2014). Design strategies in different narrative frames. Design Issues, 30 (1), 19–35. https://doi.org/10.1162/DESI_a_00246

Green, M. C., & Bavelier, D. (2012). Learning and attention in action games. Current Biology, 22(6), R198–R203.

Wixon, D., & Hollands, J. G. (2005). Scenario-Based Design: A Human-Centred Approach to Information System Development. John Wiley & Sons.

Hassenzahl, M., & Tractinsky, N. (2006). User experience–a research agenda. Behaviour & Information Technology, 25(2), 91–97.

Nielsen, J. (1993). Usability Engineering. Academic Press.

Gaver, W. W., & Schwartz, D. L. (Eds.). (2003). Principles of Learning and Instruction: Theory and Practice. Academic Press.

Carroll, J. M. (Ed.). (1995). Scenario-Based Design: Envisioning Work and Technology in System Development. John Wiley and Sons.

Simon, H. A. (1969). The Sciences of the Artificial. The MIT Press.

Aarseth, E. (1997). Cybertext: Perspectives on Ergodic Literature. Johns Hopkins University Press.

Vervoort, J. M., Bendor, R., Kelliher, A., Strik, O., & Helfgott, A. E. R. (2015). Scenarios and the art of worldmaking, Futures, 74, 62–70. https://doi.org/10.1016/j.futures.2015.08.009.

Stackelberg, P., & Alex, M. (2015). What in the world? Storyworlds, science fiction, and futures studies. Journal of Futures Studies, 20(2), 25–46.

Zaidi, L. (2017). Building Brave New Worlds: Science Fiction and Transition Design. Accessed February 20, 2023. https://openresearch.ocadu.ca/id/eprint/2123/

Raven, P. H., & Elahi, S. (2015). The new narrative: applying narratology to the shaping of futures outputs. Futures, 74, 49–61.

Negroponte, N. (1995). Being Digital. Alfred A. Knopf.

Ramirez, R., Selsky, J. W., & Van der Heijden, K. (2010). Business planning for turbulent times: new methods for applying scenarios. Routledge.

10 Digital Curation and Evaluation

Affective Evaluation Strategies under the Perspective of Human-AI Interaction

Jiayue Wang

EXPERIENTIAL VALUE IN HUMAN-AI INTERACTION

In recent years, the rapid advances in artificial intelligence (AI) technologies have significantly transformed human-computer interaction (HCI) by embedding AI systems more deeply into various aspects of our daily lives. For instance, multimodal AI assistants like GPT-4 have demonstrated remarkable capabilities in understanding and generating text, images, and even code, enabling more natural and versatile human-AI interactions (Achiam et al., 2023). In healthcare, AI-powered systems are being used to analyze medical images, assist in diagnoses, and even predict patient outcomes, fundamentally changing how healthcare professionals interact with technology (Esteva et al., 2021).

These developments have led to new challenges and trends in human-computer interaction, particularly in the realm of affective computing and emotional AI. As AI systems become more sophisticated, there is a growing focus on developing emotionally intelligent interfaces that can recognize, interpret, and respond to human emotions. Recent advancements in computer vision and natural language processing have enabled more accurate emotion recognition from facial expressions, voice, and text (Zhao et al., 2023). For example, researchers have developed AI models that can detect emotions in real-time during video calls, potentially enhancing remote communication and telehealth services (Hickson et al., 2019).

Looking ahead, the trend towards more natural and emotionally aware human-AI interactions is likely to continue. Researchers are exploring the potential of embodied AI agents and social robots that can engage in more nuanced, context-aware conversations and even display their own simulated emotions. For instance, recent studies have investigated the use of socially assistive robots in eldercare, demonstrating their potential to provide companionship and cognitive stimulation (Giorgi et al., 2022). This evolution in AI capabilities raises important questions about the nature of human-machine relationships and the ethical implications of emotionally manipulative AI systems. Scholars are actively debating the psychological impact of

DOI: 10.1201/9781032693606-13

AI companionship and the potential risks of over-reliance on AI for emotional support (Luxton, 2020).

Moreover, the integration of AI in augmented and virtual reality environments is opening up new frontiers in human-computer interaction. Researchers are developing AI-powered virtual humans that can serve as trainers, therapists, or social companions in immersive environments, potentially revolutionizing fields such as education, mental health, and social skills training (Kim et al., 2018).

Human-Computer Interaction (HCI) has been providing guidelines, principles, strategies, and recommendations for designing these systems with a "human-centered" approach (Lazar et al., 2017). When AI is injected into computer and information systems, new design challenges and opportunities arise. "Human-AI Interaction" (HAII) has thus become a cutting-edge topic in the field of HCI research (Amershi et al., 2019). AI technologies have evolved to become key components of our digital ecosystems, ranging from virtual assistants, chatbots to self-driving vehicles and smart home devices. With the continued progression and increasing complexity of AI, we must not only pay attention to its functional progress but also emphasize the quality of human experience in HAII.

In the field of HCI, one notable concept is experiential value. Experiential value refers to the overall subjective experience and satisfaction generated from interacting with AI systems (Peeters et al., 2021). It encompasses elements like enjoyment, involvement, aesthetic reception, and emotional responses, going beyond the singular functionality of human-computer or human-AI interaction (Pullman & Gross, 2004).

In this section, we conduct an in-depth examination of experiential value in the field of HCI. Academic traditions discussed here all have long histories and rich theories and methods. However, our focus is on value exploration and its practical applications in AI technologies.

First, human-AI value alignment advocates understanding and adapting to human sentiments through AI technologies while designing and optimizing AI systems based on human emotional feedback (Gabriel, 2020). Research and practices in this area aim to achieve optimal communication between humans and machines by maximizing AI systems' perception, recognition, and understanding of human emotions, and enabling intelligent, sensitive, and friendly responses (Picard, 2010; Schuller, 2018).

Second, experiential value refers to considering AI systems' holistic and overall impacts on users from a user-centered perspective when designing them. This includes utilizing AI technologies to enhance experience in usage, operations, emotion, and values. Research has shown that AI systems designed with a focus on user experience can lead to increased user satisfaction and adoption rates (Birjali et al., 2021). Moreover, the integration of AI in experiential design has the potential to create more personalized and adaptive user interfaces, catering to individual preferences and needs (Acheampong et al., 2020).

Last, applications of AI technologies in interaction design are introduced. Artists and designers in this area study challenges in transparency and collaboration between humans and machines. For example, arts can mediate between digital technologies and human understanding to overcome AI systems' "black box" vagueness, making AI reasoning transparent and decipherable, and improving critical thinking, human-machine relations, and long-term development awareness through design (Morris & Keltner, 2000).

In these three parts, value exploration in human-AI interactive experience is a core concept involving understanding, expressing, and realizing human values. By applying value exploration to AI-powered artistic designs and demonstrations, we can create AI systems that are more useful, efficient, and responsive to user needs. By analyzing users' perceptions, emotions, and behaviors during interaction with AI systems, we aim to reveal factors influencing experiential value and determine strategies to enhance overall user experience for more engaging HAII designs.

HOW TO MEASURE THE EXPERIENCE IN HUMAN-AI INTERACTION

The Application of Affective Computing in Experience Evaluation

The Emergence and Development of Affective Computing

Over 40 years ago, Herbert Simon emphasized the need to incorporate affective influences into problem solving, as the recognition and expression of emotion is essential for communication and understanding, constituting one of the fundamental human psychological needs. Human cognition, behavior and social interactions are driven and impacted by emotion (Salovey & Pizarro, 2003). In interpersonal exchanges, emotional communication is often utilized to convey one's intentions. Therefore, the ability to recognize, analyze, understand, and express emotion should also become an indispensable functionality for intelligent machines, as is being explored in the field of HCI research.

In the early 1990s, Professor Peter Salovey first proposed the concept of emotional intelligence and launched a series of relevant studies (Goleman, 2021). This concept was later developed into emotional quotient (EQ) by Daniel J. Goleman, compared to intelligence quotient (IQ) (Picard, 2000). Rosalind W. Picard is one of the early academics who systematically studied affective computing (Lisetti, 1998). In 1995, she proposed creating a computer system capable of perceiving, recognizing, and understanding human emotions through identifying bodily emotional signals. In 1997, she published the book Affective Computing, which defined affective computing and systematically introduced relevant research (Harada, 2002).

Kansei Engineering, an indigenous research field in Japan, aims to improve quality of life through comfort and enjoyment through better user interfaces in products and services. Akira Harada led the Construction of the Sensory Evaluation Framing Model (Cabanac, 2002),a large-scale special research project at the University of Tsukuba, bringing together experts from various fields including industrial design, robotics engineering, control engineering, information engineering, information management, cognitive science, aesthetics, and arts. The project set up robots with cameras in several important art museums worldwide, allowing remote participants to remotely control the robots via the internet to view artworks from distant locations. The manipulation process of each participant was recorded and analyzed to understand the psychological mechanisms underlying viewers' sensory evaluations of artworks.

General Concepts of Emotion

Emotion refers to one's attitude experience toward objective things characterized by different states such as joy, anger, sadness, etc. that arise from cognitive, evaluative, and responsive processes. Emotion exerts profound impacts on human behavior and decision-making, and constitutes a key factor driving social and cultural progress. The term emotion derives from the Greek word "pathos," referring to the feelings of pity aroused by tragedy (Garrison, 2003). Darwin considered emotion to be natural impulses and intuitions governed by biological principles, in contrast to reason which stems from civilization (Colman, 2015). The Dictionary of Psychology defines emotion as "attitudes experienced by an individual toward objective things that satisfy or fail to satisfy their needs" (Damasio, 2000). Antonio Damasio differentiated primary and secondary emotions (Russell, 2003), while James Russell constructed emotion from neurological states and actions (Scherer, 2000).

Though the complexity of emotion precludes a unified definition, it remains an important area of research in fields like psychology and AI. Emotion encompasses three elements (Can et al., 2023):

1. Subjective experience: One's self-perception of different emotional states.
2. External expression: Facial expressions formed by muscular actions in the body during emotional episodes, including facial, postural, and vocal expressions.
3. Physiological arousal: Bodily reactions elicited by emotion characterized by activation levels and distinct patterns.

Emotion can be explored from physiological and psychological perspectives (Bota et al., 2019). Psychologically, researchers process information from sensors to deduce emotional states in artificially intelligent machines.

Picard pioneered affective computing based on processing human emotional physiological signals, achieving progress and expanding applications. However, not all scholars agree with Picard's stance. Adopting phenomenological theories, Leahu (Leahu & Sengers, 2014), Gaver (Helander, 2014), and Hook (Höök, 2002) argue that emotion is crucial in human-human and human-machine interactions. They advocate interactive methods emphasizing emotional construction, human determination rather than traditional cognitive and biological perspectives. This approach focuses on experiences reflecting emotion and attempts to adjust responses.

In summary, ongoing developments expand and refine our understanding of emotion from diverse perspectives, providing deeper insight and approaches.

Definition of Emotion Computing

The idea of endowing computers with emotional capabilities is not novel, emerging nearly concurrently with the term "robotics." As early as 1985, one of the founders of AI, Marvin Minsky, a pioneer in AI who co-founded the MIT Artificial Intelligence Laboratory, pointed out that the challenge lay not in whether intelligent machines could have emotions, but rather if emotionless machines could achieve intelligence

(Minsky, 1991). However, at the time the notion of attributing human-like emotions to computers or robots remained largely confined to science fiction, with few academics exploring this area.

It was not until 1995 when Picard first proposed the concept of affective computing and formally established the theoretical framework in her 1997 publication "Affective Computing." She put forth that affective computing aimed to measure and analyze external emotional expressions in order to impact human-machine interaction. The thinking was to lend computers the ability to recognize and express emotion, thereby rendering human-computer engagement more natural.

The objective of affective computing research is to develop systems capable of perceiving, identifying, and comprehending human affect, and providing responsive, nuanced, and amicable feedback based on emotional states. This represents a complex endeavor influenced by myriad factors, including time, place, environment, and participants, requiring consideration of affective expressions through means such as facial expressions, verbal communication, and body language. In HCI, the machine must capture key information, discern emotional fluctuations, and react appropriately, for instance, by establishing user models to identify affect and selecting suitable models for information presentation. It also needs to offer timely feedback on current operations while forming new expectations based on emotional variations to proactively provide needed information to the user.

An affective computing group at MIT engineered the world's first wearable device for affective computing, a socio-emotional prosthetic capable of real-time detection of emotions in autistic children (Picard & Scheirer, 2001). By analyzing facial expressions and head movements, the device deduces cognitive affective states, while an instrument called "galvactivator" gauges excitement levels through skin conductance measurements. Such data serves to evaluate engagement and interest during interactions. Resembling a glove, it maps physiological arousal through light-emitting diodes, clearly depicting cognitive affective levels.

Emotion Feature Extraction and Recognition

Based on prior discussions, we now have an understanding of what emotion entails. The aim of affective computing is to quantitatively express the ambiguous nature of emotion, first requiring the classification of emotional features. The two most widely applied theoretical models for emotional classification are discrete emotion models and dimensional emotion models.

Discrete emotion models consider emotions as independent labels lacking associations with one another. American psychologist Carroll Izard established a basic emotion classification model of 11 discrete emotions through factor analysis, including interest, surprise, distress, disgust, enjoyment, anger, fear, sadness, shyness, contempt, and guilt (Izard, 2013). Ekman further derived a more universally accepted basic emotion classification model of seven emotions – happiness, sadness, anger, disgust, surprise, fear, and contempt – through facial expression analysis. Discrete emotion models better reflect folk psychology and everyday expression, capturing fundamental human emotional types with clear interpretability (Ekman, 1999).

Dimensional emotion models represent emotions as vectors across multiple dimensions of affective space. In the two-dimensional classification model, a notable

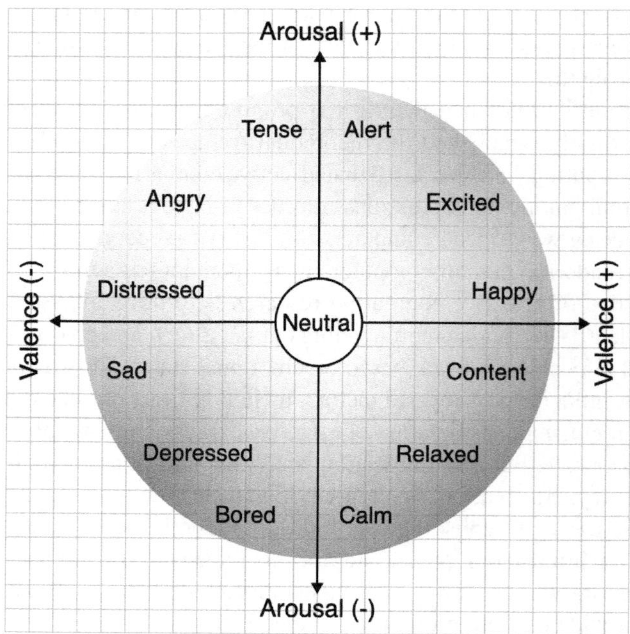

FIGURE 10.1 The Circumplex Model of Emotions. (Adapted from Russell and Barrett (1999).)[1]

instance is James Russell's Circumplex Model of Affect (Figure 10.1), known also as the VA Emotion Model for its valence-arousal structural axes (Russell & Barrett, 1999). Additional examples encompass three-dimensional emotion models typically delineating emotion types via axes and polar points, such as the pleasure-arousal-dominance model and pleasure-intensity-attention model composed of these three dimensions. Also notable is Robert Plutchik's "Wheel of Emotions" model built upon an evolutionary theory of emotion, referred to as the cone-shaped three-dimensional emotion model.

These models not only provide deeper understanding of emotion but also serve as important theoretical foundations for affective computing. Based on this groundwork, researchers attempt to quantify emotion into characterizable data in driving developments in HCI and experiential research. Continued efforts aim to empirically measure and computationally represent the dynamics of human affect.

Collection and Analysis of Emotional Signals

There are several ways to collect emotion data. One method is through the use of self-reporting tools, such as daily questionnaires and emoticons associated with annotations in documents (Hemment et al., 2023). Another method is through the use of mobile applications installed on smartphone devices, which allow for data collection in real-time and can be accessed by researchers through a web application (Seo et al., 2020). Additionally, data can be collected through the use of multimodal approaches, which involve gathering and processing data from different sources, such as electroencephalogram (EEG) devices, to support emotion detection and recognition (Lavoué et al., 2017). These methods provide researchers with valuable insights into the emotional experiences of individuals in various contexts, including distance learning, MOOCs, and educational scenarios (Lyall et al., 2016; Salmeron-Majadas et al., 2013).

Emotion Data Analysis and Visualization

Text-Based Sentiment Analysis

Text serves as a medium for human communication and information transfer. Sentiment classification is fundamental to text-based sentiment analysis, traditionally approached through five main methods: constructing sentiment lexicons, comparative analysis based on machine learning, combining sentiment lexicons and machine learning, weak supervision, and deep learning.

Supervised, semi-supervised, and unsupervised machine learning constitute three classic paradigms. Supervised learning entails classification through training optimized models to map inputs to outputs for simple judgments. However, traditional machine learning-based methods face limitations in handling diverse, complex texts, often relying on hand-crafted features and rules incapable of capturing subtle sentiment variations and semantic contexts. Performance also depends on training data scale and quality.

The emergence of generative pre-trained transformers like GPT positively impacted text-based sentiment analysis. Through self-supervised pre-training on vast corpora, GPT acquires rich linguistic representations and semantic knowledge facilitating better understanding and modeling of sentiment contents including vocabularies, intensities, and contextual associations. GPT's generative capabilities also provide new analytical perspectives – instead of classification, it can generate coherent, emotionally tinged texts exploring sentiment through inference.

GPT's transfer learning further enhances analysis. Fine-tuning on sentiment-annotated data customizes GPT for specific tasks, improving classification accuracy and generalizability. Such methods productively synergize GPT's pre-trained knowledge and sentiment comprehension with conventional techniques, achieving better results while promoting applications in sentiment intelligence, social media mining, and public opinion monitoring to advance human sentiment understanding.

Multimodal Sentiment Analysis

While text can independently convey sentiment, human communication relies more on holistic information integration. Accordingly, multimodal analysis better accords with human perception and expression of affect. Results show it outperforms unimodal text-based analysis. Emerging research focuses extensively investigates two modal combinations: audio-text and video-text analysis.

Traditional sentiment computation primarily utilizes single modal data like text, speech, expressions, gestures, or biosignals for analysis and recognition. However, human emotion expression often involves joint representation through multiple means, limiting unimodal approaches. Due to emotion's richness and diversity, integrating information from different sources through coordinated optimization can optimize recognition. Multimodal fusion algorithms consolidate representations from distinct modalities into a stable representation, effectively addressing this challenge. Commonly used fusion methods by fusion stage include early feature-level, mixed model-level, and late decision-level fusion.

Voice Emotion Computing

While speech-based sentiment analysis shows great promise, the field remains nascent with many challenges. Current issues include lack of widely accepted datasets, difficulties in annotation, and unclear relationships between acoustic features and emotion mappings.

Speech sentiment recognition aims to identify the affective state latent in speech signals beyond lexical accuracy, mining intrinsic emotional qualities conveyed through prosody and intonation variations. With advances in speech recognition, speech-based sentiment analysis holds increasing value for HCI applications. Traditional methods involve statistical and discriminative classifiers like HMM, GMM, KNN, ANN, decision trees, and SVM, though limited for complex sentiment tasks.

Deep learning architectures and high performance render neural models central to speech sentiment recognition. Various deep models including DBM, RNN, CNN, LSTM, and attention-based LSTM have been extensively applied. Notably, GPT's powerful natural language processing capabilities open new opportunities through integrating its language understanding and generation with acoustic feature extraction.

Applications span online conferences, social media, and advertising where GPT's transfer learning can apply insights from text sentiment analysis to real-time speech tone and sentiment monitoring, better informing businesses. By analyzing emotional visual qualities, GPT could also evaluate image sentiment polarity and intensity to provide consumer insight.

While holding great application potential, challenges remain due to lack of standardized datasets, annotation difficulties, and unclear relationships between acoustic correlates and affective states. Future work aims to resolve such issues through multimodal approaches leveraging visual, linguistic, and paralinguistic cues to advance the field.

Visual-Based Sentiment Analysis

In the era of social media, the proliferation of camera-enabled mobile devices has generated vast amounts of images and videos disseminated across various online and offline exhibitions and events. Researchers attempt employing suitable models to identify the affective information conveyed through these visual contents. Current research focuses include facial expression-based and bodily gesture-based sentiment recognition.

Facial expression-based approaches extract affective cues from variations in facial muscle movements and appearances. Compared to faces, bodily gestures allow for freer expression of more complex emotions and intents. This endows machines with capacity to discern richer, subtler sentiments and inner states beneath surfaces.

Deep learning has achieved notable success across diverse domains, including computer vision tasks like image classification, recognition, and retrieval. Visual sentiment analysis leveraging deep models demonstrates higher robustness and accuracy compared to traditional methods, seeing widespread application.

GPT's natural language generation capabilities provide textual interpretation and description for visual sentiment analysis. Emotions in images or videos typically require linguistic explanation. Whereas conventional methods rely on low-level feature extraction, GPT can produce descriptive text related to visual content for richer, more precise affective insights.

Additionally, GPT's generative ability presents a novel analytical approach – instead of classification, it can generate coherent emotionally tinged narratives. This opens opportunities for generative visual sentiment analysis to explore and infer emotions through generated text, facilitating more comprehensive understanding.

Physiological Signal-Based Sentiment Analysis

Widespread use of low-cost, portable high-precision sensors has rapidly advanced affective computing leveraging physiological indicators. Broadly defined, all bodily changes can index affect. Commonly utilized physiological features in sentiment research include electroencephalography, heart rate/variability, and galvanic skin response (Ahmad & Khan, 2022).

Research integrating affective recognition with consumer and exhibition contexts to enhance interaction experiences continues, seeing broad application in psychology/therapy and security as the field advances. While innovations and applications continue emerging with technological improvements, challenges remain regarding sensor constraints, signal variability, imprecise labeling, data processing difficulties, privacy concerns, and non-affective physiological influences on signals. Overcoming such hurdles through multimodal fusion holds promise to significantly advance the field.

Visualization techniques can be used to better understand emotion data. One approach is to generate time series visualizations of emotions, which can help stakeholders understand the underlying trends behind events or stories.

The work analyzes emotional expressions related to COVID-19 on Twitter (Laura-Ochoa & Tejada-Toledo, 2020). It efficiently processes massive tweets and reveals emotions through innovative tools. The tool not only helps experts gain insight into user profiles and daily tweets, but also provides a new perspective on human emotions. In preprocessing, redundant words were removed for accurate text analysis. Advanced RNNs determine tweet emotions, and emotion, arousal, and dominance scores reveal inherent emotions. The tool visually presents data over time through word clouds and two-dimensional graphs of words and scores. Its innovative features uniquely support deep exploration of emotion nuances in the analysis field.

Ma et al. developed a new approach to present emotions for everyday users in two-dimensional geography by fusing spatiotemporal information with sentiment data (Ma et al., 2020). They developed EmotionDisc, an effective tool for collecting audience sentiment based on a sentiment expression model.

KANSEI EVALUATION IN AFFECTIVE ENGINEERING'S HUMAN-INTELLIGENCE INTERACTIVE EXPERIENCE

Definition of Emotion and sensibility (Kansei/Affective Engineering)

Emotion

Feelings that occur as a result of certain experiences, such as joy, anger, pathos, humor, and like or dislike. "Emotions" are felt only by people with the sensitivity toward incentives from certain circumstances.

Sensitivity (Kansei in Japanese)

In Kant's philosophy, sensibility is distinct from reason and enlightenment. It is the capacity to receive impressions from the external world, providing the raw material for cognition and understanding.

Sensibility can be defined as the cognitive ability and wisdom to judge and comprehend things. In Kantian philosophy, sensibility is considered a necessary capacity for the exercise of thinking and intellect. While many people perceive thinking and intellect as superior abilities to sensibility, it is surprising that sensibility actually guides these abilities.

Furthermore, in her book "Perception and Sensibility," psychologist Kayo Miura provides the following definitions of sensibility (Miura, 2011):

a. Based on ambiguous and incomplete information.
b. Integrates multiple pieces of information (within sensibility).
c. Operates unconsciously and without conscious processes.
d. Instantaneous and intuitive.
e. Takes the form of impressions and evaluations.
f. Capable of making judgments that are honest and insightful, in accordance with actual circumstances.
g. Possesses the potential for learning, without being opposed to intellect, knowledge, or reason.
h. Exhibits active abilities such as creativity and discovery.

These definitions highlight sensibility as an ability that processes ambiguous information, synthesizes multiple pieces of information, and makes rapid intuitive judgments without conscious processing. Sensibility also has the potential for learning, is not opposed to intellect, knowledge, or reason, and possesses active qualities such as creativity and discovery.

Kansei is a Japanese term used to express people's impression of products, situations, and environments. It is deeply ingrained in Japanese culture, and its meaning includes sensitivity, sentimentality, emotions, and moods. From a psychological perspective, Kansei is the harmonious unity of knowledge, emotions, and sentiments. Kansei can be indirectly measured by measuring sensory activities, internal factors, psychophysiology, and behavioral responses.

MEASURING EMOTIONAL EXPERIENCE IN DESIGN – KANSEI ENGINEERING

Kansei engineering is a methodology for emotional product development that involves acquiring emotional knowledge, defining emotional goals, and generating emotional ideas (Yan, 2023). The term Kansei does not have a direct translation in any western language, and its definition in the literature is imprecise (Levy & Yamanaka, 2006). Kansei engineering is a method for converting user emotions into production parameters, and it is important in designing user interfaces to increase user satisfaction (Sobue et al., 2008). Kansei information processing aims to simulate and recognize human sensibility, sensuality, or emotion, and is used to achieve a human-centered world (Lokman, 2010; López et al., 2021). Kansei Engineering is a framework that can be used to capture consumers' implicit needs, including emotional experiences, in product design. The phase consists of four stages, as shown in Figure 10.2.

MEASUREMENT OF KANSEI

Kansei measurement is the process of capturing consumers' Kansei. Since Kansei is subjective, ambiguous, and unstructured, it is impossible to measure it directly. Therefore, we need to devise indirect measurement methods by using alternative expression approach. Kansei measurement is classified to physiological measures and psychological measures. Physiological measure targets to capture consumer behaviors, response, and body expressions. This can be done by means of analysis of brain waves by EEG, muscular loads measurement by electromyography (EMG), eye movement, and other physiological ergonomic indicators which are used to measure Kansei while a consumer is using or looking at the product. Example of study performed using this kind of measure can be found in the impact of heat to heart rate, refrigerator design, and response to robot movement.

Psychological measure deals with human mental state such as consumer behavior, expression, action, and impression. This can be measured using self-reporting system such as Different Emotional Scale (DES), Semantic Differential (SD) scale, or free labeling system. Evaluation of Kansei gives opportunity for one study to investigate the similar meanings, structure, and the concept in consumer Kansei.

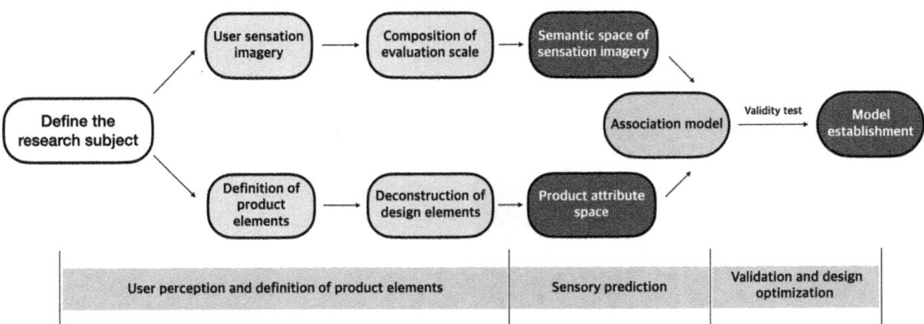

FIGURE 10.2 The phase of Kansei Engineering.

HUMAN-AI INTERACTION DESIGN BASED ON PERCEPTUAL EVALUATION

Emotional experience in HAII refers to the role of emotions in influencing cognitive processing, social interactions, and the development of human-machine interaction. Emotions play a critical role in motivation, cognitive processes, and social interactions (Brave & Nass, 2007). Designers and developers can be seen as mediators of social experience, and understanding the immaterial dimensions of technology integration in human societies is important in human-technology interaction (Rousi & Alanen, 2021). AI agents, such as virtual assistants, have the potential to adapt to users' emotional states, but currently, there is a lack of theoretical models for affective interactions with AI assistants (Colombo et al., 2021). Emotional intelligence in AI systems can lead to refined decision-making and has real-life applications (Ranade et al., 2018). Human-robot interaction can also have emotional consequences, as beliefs about and interactions with robots can change emotional concepts associated with well-being, performance, and interaction style (Weis & Herbert, 2022).

The Babypapa robot developed by Auglab[2] has a high degree of intelligence and is designed specifically for interacting with children. This robot can not only capture children's smiles and precious moments, but also permanently preserve these beautiful moments through taking pictures (Figure 10.3). In addition, the Babypapa robot has deepened the emotional connection between family members through

Concept

The robots interact with the children, capture smiles and casual moments in photos, and create opportunities to enhance relationships between family members.

FIGURE 10.3 Babypapa robots.[3]

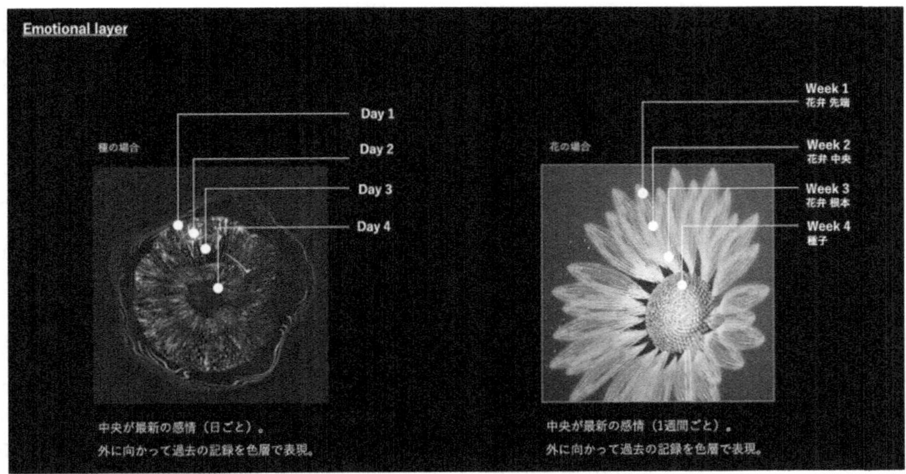

FIGURE 10.4 Concept of Log Flower.[4]

innovative AI technology, creating favorable conditions for establishing closer family relationships.

For children, the three Babypapa robots can communicate with them in a non-verbal way and bring them joyful singing and dancing performances. For parents, these robots can take photos of their children's daily lives from three different perspectives, providing them with richer recording methods.

With the continuous advancement of machine learning technology, Babypapa robot will continue to improve its intelligence level and better adapt to the needs of children of different ages and various family environments.

Tokyu Agency and Konel have co-developed a new digital plant called "Log Flower" (Figures 10.4 and 10.5). This unique plant grows by nurturing the atmosphere

FIGURE 10.5 Test scenario for collecting physiological signals at an exhibition venue.[5]

of the room and functions as a "mirror of space." It changes its growth pattern based on the emotions it absorbs. "Log Flower" is the only plant of its kind in the world that makes people aware of current communication styles and aims to cultivate positive mindsets.

CONCLUSION

In the AI era, the design of affective HCIs is undergoing transformative changes in many aspects, bringing new trends and challenges in interaction modalities, presentation interfaces, language exchange, and emotional perception. From the perspective of interaction modalities, design is shifting from unimodal to multimodal interactions, incorporating visual, auditory, and haptic experiences to realize more natural and human-centered HCIs.

In terms of presentation interfaces, the trend is toward the convergence of physical and digital realities, moving from simple flat displays to immersive and high-fidelity virtual environments that provide users with more engaging and emotional experiences.

For interaction language, there is an evolution from command-based voice interactions to natural language engagement, with the goal of creating a deeper emotional connection through more human conversational exchange.

The core challenge in AI affective design lies in emotional perception, which requires the integration of multimodal sensing and machine learning technologies to accurately detect and appropriately respond to human emotions, thereby enhancing the emotional intelligence of HCI.

Alongside these overarching trends in the AI era, this chapter delineates four distinct trajectories positioned along two pivotal axes: the "AI Affective-Human Affective" axis of sensibility and the "Computing-Expressing" axis of interaction process (Figure 10.6).

First, concerning the quadrant encapsulating "Human Affective" and "Computing," there is an evident trend toward modeling the intrinsic emotional states of users. Advanced measurement techniques employing physiological, facial, vocal, and behavioral analytics permit a more nuanced apprehension of user sentiments.

Second, transitioning from "Human Affective" to "Expressing," the translation of these intrinsic emotional cues into machinable interactions has gained traction. The adaptive interfaces reflect empathetic engagement, dynamically altering visual and auditory elements to resonate with the user's affective state.

Third, the exploration into "AI affective" combined with "Computing" illustrates a nascent yet burgeoning field. Here, the focus is to fabricate an affective construct within AI systems. New algorithms are being designed to not only measure but also to contextualize the emotional output generated by AI, effectively bridging the gap between human emotions and AI responses.

Finally, in the realm of "AI affective" and "Expressing," there is a deliberate push toward enabling AI systems to exhibit emotive communication. Such systems are engineered to simulate emotional intelligence, adjusting their interaction nuances based on the conversational matrix and environmental stimuli, thus displaying a spectrum of affective responses that mimic human-like expressiveness.

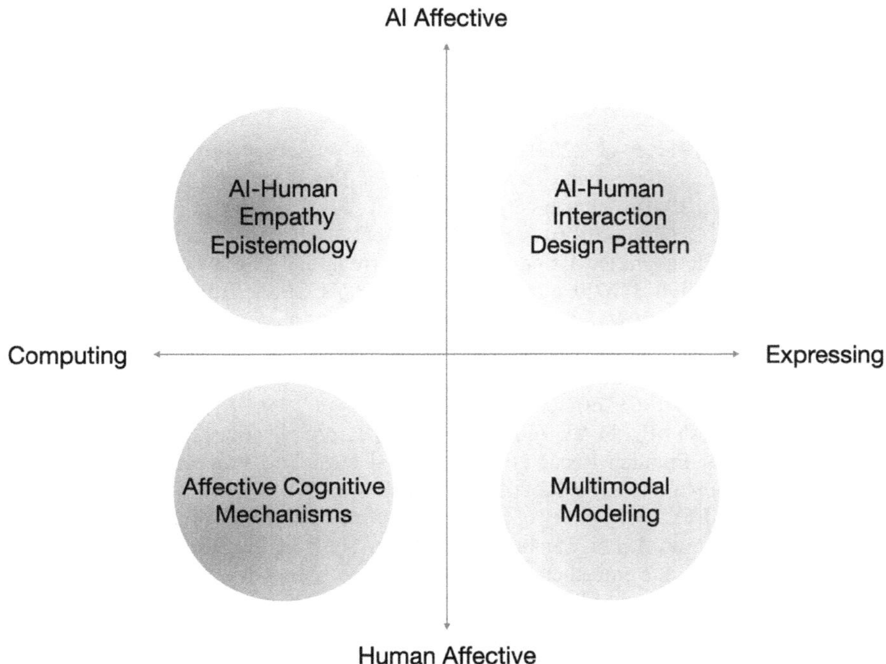

FIGURE 10.6 The Trend Framework for emotional human-AI interaction.

In conclusion, these emergent trends highlight a transformative phase in affective interaction design, where the emotional interplay between humans and AI is increasingly sophisticated. As AI continues to evolve, the implications of these advancements suggest a future where the emotional granularity of HAII is profoundly enhanced, fostering more intuitive and empathetic user experiences.

NOTES

1 Circumplex Model of Emotion, From Wikimedia Commons: https://commons.wikimedia.org/wiki/File:Circumplex_model_of_emotion.svg
2 Auglab: https://tech.panasonic.com/global/auglab/ideas/babypapa.html
3 Babypapa robots: https://tech.panasonic.com/global/auglab/ideas/babypapa.html
4 Log Flower: https://chizaizukan.com/property/748/
5 Log Flower: https://chizaizukan.com/property/748/

REFERENCES

Acheampong, F. A., Wenyu, C., et al. (2020). Text-Based Emotion Detection: Advances, Challenges, and Opportunities. Engineering Reports, 2(7), Article e12189. https://doi.org/10.1002/eng2.12189

Achiam, J., Adler, S., et al. (2023). Gpt-4 Technical Report. arXiv. https://arxiv.org/abs/2303.08774

Ahmad, Z., & Khan, N. (2022). A Survey on Physiological Signal-Based Emotion Recognition. Bioengineering, 9(11), 688.

Amershi, S., Weld, D., et al. (2019, May). Guidelines for Human-AI Interaction. In Proceedings of the 2019 Chi Conference on Human Factors in Computing Systems, 1–13.

Birjali, M., Kasri, M., et al. (2021). A Comprehensive Survey on Sentiment Analysis: Approaches, Challenges and Trends. Knowledge-Based Systems, 226, Article 107134. https://doi.org/10.1016/j.knosys.2021.107134

Bota, P. J., Wang, C., et al. (2019). A Review, Current Challenges, and Future Possibilities on Emotion Recognition Using Machine Learning and Physiological Signals. IEEE Access, 7, 140990–141020. https://doi.org/10.1109/ACCESS.2019.2944001

Brave, S., & Nass, C. (2007). Emotion in human-computer interaction. In The human-computer interaction handbook (pp. 103–118). CRC Press.

Cabanac, M. (2002). What is Emotion? Behavioural Processes, 60(2), 69–83. https://doi.org/10.1016/S0376-6357(02)00078-5

Can, Y. S., Mahesh, B., et al. (2023). Approaches, Applications, and Challenges in Physiological Emotion Recognition-A Tutorial Overview. Proceedings of the IEEE. https://doi.org/10.1109/JPROC.2023.3286445

Colman, A. M. (2015). A dictionary of psychology (4th ed.). Oxford University Press.

Colombo, S., Rampino, L., & Zambrelli, F. (2021). The adaptive affective loop: how AI agents can generate empathetic systemic experiences. In Advances in Information and Communication: Proceedings of the 2021 Future of Information and Communication Conference (FICC), Volume 1 (pp. 547–559). Springer International Publishing.

Damasio, A. R. (2000). A second chance for emotion. In R. D. Lane & L. Nadel (Eds.), Cognitive neuroscience of emotion (pp. 12–23). Oxford University Press.

Ekman, P. (1999). Facial expressions. In T. Dalgleish & M. J. Power (Eds.), Handbook of cognition and emotion (pp. 301–320). New York, NY: Wiley.

Esteva, A., Chou, K., et al. (2021). Deep Learning-Enabled Medical Computer Vision. NPJ Digital Medicine, 4(1), 5. https://doi.org/10.1038/s41746-020-00376-2

Gabriel, I. (2020). Artificial Intelligence, Values, and Alignment. Minds and Machines, 30(3), 411–437. https://doi.org/10.1007/s11023-020-09539-2

Garrison, J. (2003). Dewey's Theory of Emotions: The Unity of Thought and Emotion in Naturalistic Functional "Co-ordination" of Behavior. Transactions of the Charles S. Peirce Society, 39(3), 405–443.

Giorgi, I., Tirotto, F. A., et al. (2022). Friendly but Faulty: A Pilot Study on the Perceived trust of Older Adults in a Social Robot. IEEE Access, 10, 92084–92096. https://doi.org/10.1109/ACCESS.2022.3202942.

Goleman, D. (2021). Leadership: The power of emotional intelligence. More Than Sound LLC.

Harada, A. (2002). Towards Establishment of Kansei Science. Special Issue of Japanese Society for the Science of Design, 10(2), 39–46. [In Japanese]

Helander, M. G. (Ed.). (2014). Handbook of human-computer interaction (pp. 1003–1041). Elsevier, North-Holland.

Hemment, D., Murray-Rust, D., Belle, V., Aylett, R., Vidmar, M., & Broz, F. (2023). Experiential AI: A Transdisciplinary Framework for Legibility and Agency in AI. arXiv preprint arXiv:2306.00635.

Hickson, S., Dufour, N., et al. (2019). Eyemotion: Classifying Facial Expressions in VR Using Eye-Tracking Cameras. In 2019 IEEE Winter Conference on Applications of Computer Vision (WACV), 1626–1635. IEEE. https://doi.org/10.1109/WACV.2019.00178

Höök, K. (2002). *Evaluating affective interaction.* In *AAMAS'2002: Workshop on Embodied Conversational Agents - Let's Specify and Evaluate Them.* RISE, Swedish ICT, SICS. https://vhml.org/workshops/AAMAS/papers/hook.pdf

Izard, C. E. (2013). Human emotions. Springer Science & Business Media.
Kim, K., Boelling, L., et al. (2018). Does a Digital Assistant Need a Body? The Influence of Visual Embodiment and Social Behavior on the Perception of Intelligent Virtual Agents in AR. In 2018 IEEE International Symposium on Mixed and Augmented Reality (ISMAR), 105–114. IEEE. https://doi.org/10.1109/ISMAR.2018.00039
Laura-Ochoa, L., & Tejada-Toledo, F. (2020, September 16–18). CovidEmoVis - An interactive visual analytic tool for exploring emotions from Twitter data of Covid-19. In Human-computer interaction: 6th Iberomarican workshop, HCI-Collab 2020, Arequipa, Peru, Proceedings 6 (pp. 94–106). Springer International Publishing.
Lavoué, E., Molinari, G., & Trannois, M. (2017). Emotional Data Collection Using Self-Reporting Tools in Distance Learning Courses. http://doi.org/10.1109/ICALT.2017.94
Lazar, J., Feng, J. H., & Hochheiser, H. (2017). Research methods in human-computer interaction. Morgan Kaufmann.
Leahu, L., & Sengers, P. (2014). Freaky: Performing Hybrid Human-Machine Emotion. In Proceedings of the 2014 Conference on Designing Interactive Systems, 607–616. https://doi.org/10.1145/2598510.2600879
Levy, P., & Yamanaka, T. (2006). Towards a definition of Kansei. In Friedman, K., Love, T., Côrte-Real, E. and Rust, C. (Eds.), Wonderground - DRS International Conference 2006, 1–4.
Lisetti, C. L. (1998). Review of the book affective computing, by R. Picard. MIT Press.
Lokman, A. M. (2010). Design & Emotion: The Kansei Engineering Methodology. Malaysian Journal of Computing, 1(1), 1–11.
López, Ó., Murillo, C., & Gónzalez, A. (2021). State of the Art Analysis of Emotional Design Methodologies and Their Demonstrated Results. In Advances in Industrial Design: Proceedings of the AHFE 2021 Virtual Conferences on Design for Inclusion, Affective and Pleasurable Design, Interdisciplinary Practice in Industrial Design, Kansei Engineering, and Human Factors for Apparel and Textile Engineering, July 25-29, 2021, USA (pp. 943–951). Springer International Publishing.
Luxton, D. D. (2020). Ethical Implications of Conversational Agents in Global Public Health. Bulletin of the World Health Organization, 98(4), 285. https://pubmed.ncbi.nlm.nih.gov/32284654/
Lyall, S., Elmiligi, H., & Ortner, C. N. M. (2016). Using Mobile and Web Technologies to Collect and Analyze Emotion Survey Data. http://doi.org/10.1145/2910925.2910939
Ma, C., Song, J. -C., Zhu, Q., Maher, K. T., Maher, K. T., Huang, Z. -Y., & Wang, H. (2020). EmotionMap: Visual Analysis of Video Emotional Content on a Map. Journal of Computer Science and Technology, 35(3), 576–591. http://doi.org/10.1007/S11390-020-0271-2
Minsky, M. L. (1991). Logical Versus Analogical or Symbolic Versus Connectionist or Neat Versus Scruffy. AI Magazine, 12(2), 34–34.
Miura, K. (2011). Kansei as Mental Activity: Perception with Impression, Intuitive Judgment and the Basis of Creativity. Japanese Psychological Research, 53(4), 341–348. https://doi.org/10.1111/j.1468-5884.2011.00485.x
Morris, M. W., & Keltner, D. (2000). How Emotions Work: The Social Functions of Emotional Expression in Negotiations. Research in Organizational Behavior, 22, 1–50. https://doi.org/10.1016/S0191-3085(00)22002-9
Peeters, M. M., van Diggelen, J., et al. (2021). Hybrid Collective Intelligence in a Human–AI Society. AI Society, 36, 217–238.
Picard, R. W. (2000). Affective computing. MIT Press.
Picard, R. W. (2010). Affective Computing: From Laughter to IEEE. IEEE Transactions on Affective Computing, 1(1), 11–17. https://doi.org/10.1109/T-AFFC.2010.10
Picard, R. W., & Scheirer, J. (2001, August). The Galvactivator: A Glove that Senses and Communicates Skin Conductivity. In Proceedings 9th Int. Conf. on HCI.

Pullman, M. E., & Gross, M. A. (2004). Ability of Experience Design Elements to Elicit Emotions and Loyalty Behaviors. Decision Sciences, 35(3), 551–578.

Ranade, A. G., Patel, M., & Magare, A. (2018). Emotion Model for Artificial Intelligence and Their Applications. http://doi.org/10.1109/PDGC.2018.8745840

Rousi, R., & Alanen, H. -K. (2021). Socio-Emotional Experience in Human Technology Interaction Design – A Fashion Framework Proposal. http://doi.org/10.1007/978-3-030-77431-8_8

Russell, J. A. (2003). Core Affect and the Psychological Construction of Emotion. Psychological Review, 110(1), 145–172. https://doi.org/10.1037/0033-295X.110.1.145

Russell, J. A., & Barrett, L. F. (1999). Core Affect, Prototypical Emotional Episodes, and Other Things Called Emotion: Dissecting the Elephant. Journal of Personality and Social Psychology, 76(5), 805.

Salmeron-Majadas, S., Santos, O. C., Boticario, J. G., Cabestrero, R., Quirós, P., & Saneiro, M. (2013, July 6-9). *Gathering emotional data from multiple sources.* In S. K. D'Mello, R. A. Calvo, & A. Olney (Eds.), *Proceedings of the 6th International Conference on Educational Data Mining* (pp. 404–405). International Educational Data Mining Society. http://www.educationaldatamining.org/EDM2013/papers/rn_paper_100.pdf

Salovey, P., & Pizarro, D. A. (2003). The value of emotional intelligence. In R. J. Sternberg, J. Lautrey, & T. I. Lubart (Eds.), Models of intelligence (pp. 263–278). American Psychological Association.

Scherer, K. R. (2000). Psychological Models of Emotion. The Neuropsychology of Emotion, 137(3), 137–162. Oxford University Press.

Schuller, B. W. (2018). Speech Emotion Recognition: Two Decades in a Nutshell, Benchmarks, and Ongoing Trends. Communications of the ACM, 61(5), 90–99. https://doi.org/10.1145/3129340

Seo, C., Zhang, L., Kim, Y.-H., Yun, S.-J., Lee, D., Kim, Y., & Lee, J. (2021). *Electronic device and method of obtaining emotion information* (European Patent No. EP3820369A1). European Patent Office. https://patents.google.com/patent/EP3820369A1/en

Sobue, S., Huang, X., & Chen, Y. W. (2008). Mapping Functions between Image Features and KANSEI and its Application to KANSEI Based Clothing Fabric Image Retrieval. In ITC-CSCC: International Technical Conference on Circuits Systems, Computers and Communications, 705–708.

Weis, P. P., & Herbert, C. (2022). Do I Still Like Myself? Human-Robot Collaboration Entails Emotional Consequences. Computers in Human Behavior, 127, 107060.

Yan, H. (2023). Emotional Product Development: Concepts, Framework, and Methodologies. In Knowledge Technology and Systems: Toward Establishing Knowledge Systems Science (pp. 197–225). Singapore: Springer Nature Singapore.

Zhao, X., Deng, Y., et al. (2023). A Comprehensive Survey on Deep Learning for Relation Extraction: Recent Advances and New Frontiers. arXiv. https://arxiv.org/abs/2306.02051

Section IV

Future Influence

11 Design Futures Education in HCI

Peter Scupelli

INTRODUCTION

This chapter is about Design Futures in Human-Computer Interaction. "Design Futures" describes the interaction between "futures" and "design." Three assumptions. First, many may assume that "Design Futures" is limited to "Speculative Design," "Design Fiction," or "Experiential Futures." The reality is that "Design Futures" practices are expanding to include new ways to combine design and futures thinking. Second, it is possible to invent new design methods. Third, "Design Futures" is generic; it lacks directionality and specific "values." Adding directionality and values allows one to become specific with the alternative Design Futures they explore.

Futures Studies or Futures is a transdisciplinary field of inquiry investigating diverse possible, probable, and preferred futures (e.g., Bell, 2003; Gidley, 2017). In the narrowest definition, Futures explores thinking through time. Futures Studies are more expansive than that. In this chapter, I focus on several futures thinking methods.

Just about anything can be designed. Look around; design is everywhere. Notice the chair you are seated in, the vehicles you ride, the clothes you wear, the technology you use, the food you eat, the games you play, your entertainment, your glasses, and so forth. Everything is designed directly (or indirectly). So, a chapter on "Design Futures" could be about anything. In this chapter, I narrow the focus to design for sustainable futures.

Design for HCI shifts and adapts as computer-based technology evolves. Consider the trajectory in computing along the path of graphical user interfaces (GUI), groupware, mobile computing, social computing, IoT, intelligent spaces, Conversational User Interfaces (CUI), Augmented Reality (AR), Big Data, Artificial Intelligence (AI), Blockchain, Spatial Web, Metaverses, and so forth.

New forms of design emerge as HCI evolves. The field of HCI has been theorized in at least three paradigms (Harrison et al., 2007) or four waves (Bødker, 2006, 2015). The first wave focused on individual users and the human cognitive aspects of human-computer interactions. The second wave focused on human needs with groupware technologies and the context where people work together. The third wave focused on technologies that integrated from personal computers to tablets to mobile computing, leading to multiple user experiences with user-created content on online platforms and online communities. The fourth wave of HCI focuses on transdisciplinary design paradigms, including politics, values, and ethics (e.g., Lopes, 2022).

HCI is a rapidly unfolding multiverse that moves at a breakneck pace. What kind of "Design Futures for HCI" will I focus on in this chapter? I'll focus on design for 21st-century societal problems/opportunities, such as the Sustainable Development

DOI: 10.1201/9781032693606-15

Goals (SGD). The IPCC (Intergovernmental Panel Climate Change) is clear about two points: first, our current designed lifestyles are responsible for the unfolding climate disaster. Second, we must reach zero-carbon lifestyles globally by the year 2050 and reduce carbon emissions by at least 50% by the year 2030 (IPCC, 2023). Hence, I'll focus on Design Futures education that can help to focus on rapid decarbonization to accomplish zero-carbon lifestyles and other SDG-related goals.

This chapter focuses on how designers can play a critical role in helping to shift from a destructive design paradigm that "defutures" the world and takes away our collective futures to a constructive design paradigm called "design futuring" that gives us collective futures (Fry, 1999, 2008).

One big problem with the "status quo" in design thinking methods is focusing on short-term action without much concern for long-term consequences. Ignoring long-term consequences is what has created our sustainability problems. From a long-term sustainability premise, such a short-sighted perspective is unacceptable. To pursue carbon-intensive lifestyles now is to deny such actions' ethical, moral, and social consequences. We must pursue actions today that move towards desirable and sustainable futures for all.

Why shift towards a sustainability-focused design futures education? Sustainability is about doing things today that give us collective futures. Instead, the status quo short-term design paradigm focuses on designing desirable, economically viable, and technically feasible solutions. We live in dire times that require a shift from such narrow boundaries. Designers must consider their solutions' broader impacts on society and the environment, aligning their work with larger ambitions (e.g., sustainability, equity, inclusivity, anti-racism, decoloniality).

Consequently, designers must design for values that align with larger societal ambitions. Designers should keep returning to the question, what kind of future does my design help to create and sustain? How might designed products, services, and experiences move towards more desirable and sustainable futures for all?

How might we designers do that? Earlier, I wrote that the status quo in traditional design education is lacking. Many other scholars have described how design education needs to change to include an interdisciplinary approach, systems thinking, ethics, sustainability, collaboration, continuous learning, and practical experience (e.g., Fry, 2008). In this chapter, I'll focus on how futures can be mixed with design to align short-term design action with long-term sustainability vision goals.

Futures Studies has much to teach designers about how to reason seriously and rigorously about long time horizons. Designers can ensure that their short-term design actions are part of a larger road map that traces pathways to long-term future vision goals.

Some designers might object that all design is about "futures" because a designer launches a project into a previously non-existent future reality. While I agree that all projects go from present to future, the issues I raise differ.

- How far are the projects thrown temporally into the future?
- What kinds of futures are designers throwing their projects into?
- What forces of change are they aligning their projects with?
- What values and politics are their projects reinforcing or challenging?

Some readers might think that "critical design" and "socio-technical studies" answer many of these questions. I'd argue that while "critical design" and "socio-technical systems" can unpack designer's assumptions, technological perspectives, and politics, future studies provide tools to describe temporalities and paradigm shifts from a current situation to different paradigms more broadly.

What futures thinking methods are helpful for designers interested in designing futures? I'll suggest seven as a starting point: (a) explore drivers of change, (b) critique images of futures, (c) design for alternative futures, (d) deepen understandings of Futures with CLA, (e) mapping pathways linking past, present, and Futures, (f) exploring internal and external futures, and (g) structuring Design Futures experiences. In the following seven sections, I'd describe each.

EXPLORE DRIVERS OF CHANGE

What shapes our pasts, presents, and futures? Some factors that may shape our hopes and fears include social, technological, economic, environmental, and political forces. One might fear getting old in a society with limited social support for older adults. One might be enthusiastic about how technology could improve healthcare by allowing people to monitor their health. Economic forces might make it harder to export products due to the currency exchange rate. Environmental concerns might reduce the exploitation of natural resources. Political decisions can determine legal activities in a particular place or place import tariffs. Values may differ depending on the community.

This section explores how STEEP+ V forces (i.e., Social, Technology, Economics, Environmental, Political, Values) can shape what we imagine as possible futures. How might such ideas shape new products, services, and experiences? How might STEEP+V forces be embedded into design thinking processes? One can see STEEP+V forces through objective and subjective lenses. Objective lenses establish what is observable from an objective perspective. Meanwhile, subjective lenses might describe how people feel about objective realities.

Social forces include demographics (e.g., population growth, age distribution, immigration rates), lifestyles, social and cultural values (e.g., health consciousness), consumer behavior and advertising, etc. Design projects should consider the context for such social forces operating on the communities of users for products, services, and experiences.

Technological forces People interact with technological forces through new products, services, and experiences. Technological factors may include Research and Development (R&D) activity, Artificial Intelligence (AI), robotics, automation, technological incentives, and technological change. Technology often plays a central role in HCI. How might technological forces play out objectively and subjectively in designed products, services, and experiences?

Economic forces are visible at different levels and influence customers. Some examples include interest rates, taxes, international trade (e.g., exchange rates), entrepreneurship, and employment rates. Economic considerations play a core role in the viability of design solutions. The business model of a service, product, or experience plays a critical role. Larger economic forces shape the context in which business models operate.

Environmental forces such as water, wind, soil, food, energy, etc, are linked to ecosystem health. Ecological considerations for good or ill can affect the supply chain, production location choices, production processes used, and relationships between a company and local communities.

Political forces may greatly influence individuals and companies. Political forces may change dramatically after an election when different parties gain power. Political developments take place through laws. With technology companies, there is much to consider regarding political forces. For example, online election misinformation campaigns, cyberbullying, and so forth are linked to social media.

Values People can express their values as evolving preferences for social relations, leisure, entertainment, consumerism, exercise, health, social issues, etc. What attitudes describe life/work balance, work-life progression (e.g., career advancement, entrepreneurship, hierarchical structures), well-being, and so forth? Values can play a core role in determining the design direction and emphasis of a product, service, and experience.

Typically, one focuses on the external nature and context when discussing STEEP+V forces. From a design perspective, one should look deeper, though. One might notice three levels of external forces include: *Macro, Meso,* and *Micro*. First, the *Macro* Level involves global forces such as "economic development, demographics, politics, technological developments, and other external forces" (e.g., Rijn & Burgt, 2021). Designers can consider how such *macro* forces touch and interact with people's lives. What are the designed touchpoints with such forces?

Second, the *Meso* level involves a transactional level of 'market forces' such as supply/demand, distribution, competitors, and strategic alliances (Rijn & Burgt, 2021). The challenge for designers is to find strategic design opportunities within this area for designed products, services, and experiences. In other words, what might be design opportunities linked to the broader system at the *Meso* level needed to create or deliver products, services, environments, and experiences?

Third, the *Micro* level involves forces within an organization. Designers might work on the vision, mission, strategy, resources, processes, products, and services. The design opportunity is to imagine the alignments between the values embedded within the product, service, or experience and the organization itself. Designers will go further into the details of what is the interaction with and touchpoint for such products, services, and experiences.

While the *Macro*, *Meso*, and *Micro* levels are helpful in orienting, design questions typically bring such levels back to the human experiences rooted in place. How might those levels appear in people's lives, families, communities, public policy, etc? Next, I describe how one might critique differing images of possible futures.

CRITIQUE IMAGES OF FUTURES

One skill that designers develop through their training and professional work is how to critique designs and how to receive critiques. The design critique is an activity where one designer shows their work to someone else and receives comments about what the other person notices in the work. Critiques can range from focused to open-ended. A designer might ask for directed feedback on some particular design detail, or other critiques may be more open-ended.

The critique process aims to provide pointed feedback from someone else's perspective and experience. It allows the designer to see things that they might have missed. To give an honest critique is a very generous activity. It would be far easier to say that all is well and that the design is excellent. Instead, it takes courage to critique someone else's work and choices.

Receiving a critique is an art as well. It is easy to dismiss comments that question one's assumptions or design directions. The value of a critique comes from understanding why people are making their comments and what assumptions they are making. Receiving a critique takes humility, maturity, and curiosity from the designer or team.

The traditional design critique is taught tacitly through repetition. One learns to critique by carefully paying attention during design critiques and reflecting on the critiques. Next, I'll describe how designers might use their capacity to critique in the domain of Design Futures.

One of the challenges of teaching Design Futures is that, typically, there is one person (or, at best, a few) on the faculty with deep expertise in Futures Studies who can model design critiques for Design Futures-related themes. How might design students learn to critique futures work without participating in a robust futures thinking community? This section aims to provide some ways to get deeper into images of the future. First, we rely on Futurist Jamais Cascio's Bad Futurism talk (Cascio, 2012) to identify easily avoidable mistakes. Second, we provide ten questions that futurist Adam Gordon suggests one should ask about any image of the future from his book Future Savvy (2009). For deeper and more critical exploration of futures scenarios read journals like *Futures* and *The Journal of Futures Studies*.

Futurist Jamais Cascio describes ten errors people make with futures scenarios (Cascio, 2013). I've clustered the ten critiques into three types of flaws: The first flaw is focusing on partial aspects of a scenario in three ways: (a) identifying only technological advances and missing how people live; (b) assuming everything works and missing the failures and the unintended uses; and (c) focus only on the dominant class missing the broader impacts on classes in society. The second flaw involves ignoring human nature in futures scenarios. The third flaw entails a lack of respect for the audience's intelligence (e.g., make your case and trust your audience to choose the better scenario; provide equally seductive and terrifying scenarios that prepare for success and failure).

In the book *Future Savvy* by Adam Gordon (2009), there are ten questions that futurists should ask of any futures scenario. In the interest of space, I limit myself to summarizing the ten questions in one sentence. See Gordon's book to understand the critiques in more depth.

- **Purpose** – Why was the forecast made?
- **Specificity** – Is the forecast predictive or speculative?
- **Information Quality** – How good is the data?
- **Interpretation and Bias** – Are the biases natural or intentional?
- **Methods and Models** – What methods led from the present to the futures?

- **Quantitative Limits** – Are the quantitative methods appropriate?
- **Managing Complexity** – Does the forecast oversimplify the world?
- **Assumptions and Paradigm Paralysis** – Was adequate horizon scanning done?
- **Zeitgeist and Groupthink** – Is this forecast an embodiment of the Zeitgeist (signs of the times), groupthink (group consensus), or are these deep explorations into alternative futures?
- **Drivers and Blockers** – What forces are driving the changes, and what are blocking forces from such change?

In this section, we've explored the pointed critiques of Futurists Jamais Cascio and Adam Gordon on images of futures to inform as a fundamental skill designers possess. Critiques aim to surface obvious limitations and discuss ways to overcome them. Expert critics ask insightful questions and suggest helpful considerations. Just like medical doctors provide a diagnosis (what is wrong), they also give a prognosis (what one can do). Designers can use the pointed critique questions from Futurists Jamias Cascio and Adam Gordon to discover new design opportunities.

DESIGN WITHIN ALTERNATIVE FUTURES

One of the first principles of Futures Studies articulated by Jim Dator is that "the future" is fiction because "the future" does not exist (Dator, 2019). There are always multiple possible futures that we should imagine, explore, and consider (Dator, 2009). How might one go about seriously exploring multiple possible futures? There are many ways to explore alternative futures. We explore three different ways of thinking about alternative futures: Jim Dator's generic images of futures (Dator, 2009), organization-centered alternative futures based on two critical uncertainties (e.g., Schwartz, 1991) and Global Scenarios such as ARUP's Alternative Scenarios for 2050 (Schemel et al., 2019).

Jim Dator realized that no single future out there awaits our correct prediction. Many possible futures co-exist. Each possible future follows their inner logic and may contradict other futures. Dator clustered billions of existing images of futures into four generic futures: *Continued Growth, Collapse, Discipline*, and *Transformation*. *Continued Growth* is the "official" future that one hears about from governments, educational systems, and organizations. *Collapse* is another kind of future for people concerned with things worsening, either extinction or a lower standard of living and development. *Discipline* is a future where sustainability concerns overshadow "continued economic growth" to create a future with strict resource limitations. *Transformation* is a future where the transformative power of technology allows for a "transformed society" on an entirely artificial Earth and perhaps other planets (Dator, 2009).

William Gibson's famous quote, "The future is already here, just not evenly distributed," can provide deep insights into Dator's generic images of futures as lenses to read the present times (and perhaps even the past). One can see traces of these four generic futures in the past, present, and futures.

One design opportunity with the four generic images of futures is to use these different alternative worlds as backdrops to design and test current products, services, and experiences. The fit or lack of fit for a product, service, or experience is a way to stress-test a design solution in different realities. One can also use the generic images of futures to assess the present time one is designing for and imagine the pathway to a different generic future. For example, a designer may ask: what might the opportunities for a product, service, or experience in a "continued growth" world be like when, in reality, they should prepare for a different world instead (e.g., discipline, collapse, transformation)? Imagining such diverse pathways may provide opportunities to imagine strategic roadmaps for other opportunities.

2×2 ALTERNATIVE SCENARIOS

Another kind of alternative futures is created by placing two critical uncertainties on two axes to create four different scenarios. ARUP's Alternative Global Futures 2050 scenarios have two axes: planetary health (improves - worsens) and societal health (improves - worsens). Four plausible scenarios emerge *Greentocracy*, *Extinction Express*, *Humans INC.*, and *Post-Anthropocene* (Figure 11.1). *Greentocracy* is a place where planetary health improves, but societal conditions decline. *Extinction Express* is a scenario where both planetary health and societal conditions decline. *Humans INC.* is a scenario where societal conditions improve and planetary health decreases. The *Post-Anthropocene* is a scenario where both planetary health and societal conditions improve. The *Post Anthropocene* scenario aligns with successful SDG attainment (Schemel et al., 2019).

The ARUP 2050 Scenarios can help design students to imagine four plausible realities that they might design within. Such future scenarios describe reality

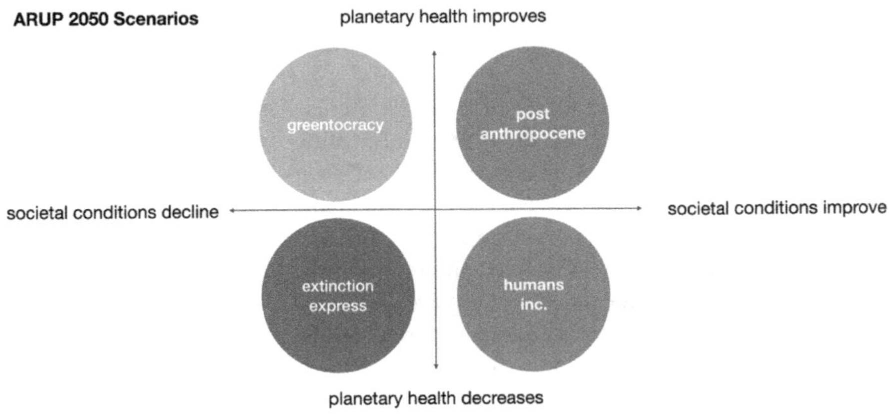

FIGURE 11.1 The Four ARUP 2050 scenarios, Greentocracy, Post-Anthropocene, Extinction Express, and Humans INC, illustrate the relationship between the two axes of planetary health and societal conditions (Redrawn image credit © LELab.)

abstractly. The design task is to imagine what it might be like to live in such a scenario. What behaviors are likely? What infrastructure is necessary to survive in such a world? What worldviews belong in such a scenario? What myths and metaphors might guide existence in such a place?

With my design students, I've used alternative scenarios in two ways: generatively for new concepts and to evaluate existing prototypes in different alternative scenarios.

Generatively, designers can craft products, services, and experiences to respond to alternative scenarios. Having alternative scenarios described broadens the design space of possibilities for the design team's imagination regarding what products, services, and experiences are possible for worlds beyond the realities they are immediately in or familiar with. As the design team explores different levels of the design space, new ideas for products, services, and experiences can emerge to meet such specific design spaces.

Futures Scenario Evaluation Likewise, an existing prototype developed for a specific place and time can be placed in alternative scenarios and tested for fit. What about a product, service, or experience would work very well? Where might new friction points emerge? Every problem identified is a potential opportunity for a new solution.

DEEPENING FUTURES

Images of Futures can quickly feel relatively superficial when focused on aspects such as technology or abstract components such as economic and political theories. It is difficult to imagine what life might feel like to inhabit such futures. What behaviors might one engage in? What infrastructure and systems would be needed? What worldviews might these future citizens hold? What stories and myths might describe their experiences? In short, a descriptive depth is required for a future experience to feel real. How might one create a detailed description of an image of the future?

Sohail Inayatullah, the inventor of Causal Layered Analysis (CLA), posits that reality has four layers (Inayatullah, 1998). Thus, future images can feel superficial when they deal only with observable behavior and focus on descriptions of technological details. There are deeper layers than behavior and infrastructure. Worldviews, the philosophy of an era, the politics of a place, and so forth add a deeper dimension to one's understanding of reality. The culture, myths, and stories that one tells of a place often encapsulate the essence of a place and times.

Design thinking tends to focus on apparent needs that people have linked to behavior. CLA analysis gives designers an understanding of reality on four layers. Thus, CLA provides designers with deep insights and opportunities left unaddressed by superficial design thinking methods. I'll argue in this section that designers can significantly benefit from using CLA to imagine multiple layers of design opportunities.

I've guided my students to use CLA to explore their "personal futures" (Scupelli, 2021, 2022), current studio projects, and the ARUP 2050 plausible scenarios for the year 2050 (Scupelli, 2023). Analyzing such diverse realities with a structured layered analysis approach afforded students deep insights into their possible personal future career options, current design studio projects, and plausible futures scenarios for the year 2050.

CLA is a valuable analysis tool to explore how one might go from a current situation to a different futures situation. First, one conducts a CLA analysis of the two time periods. Second, placing the four CLA layers side-by-side allows one to perform a gap analysis to identify necessary changes to transition from a current to a future time. Thus, CLA analysis helps designers identify potential design opportunities for transitions on each level of analysis. The CLA analysis of a specific product in the present time can guide one to imagine potential fit in a different futures image as described in the previous section on alternative futures scenarios. In the next section, I describe how one might imagine past, present, and future linkages.

MAPPING PATHWAYS

In the first section, we described how STEEP+V forces of change play a critical role in shaping context through six different lenses (i.e., social, technological, economic, environmental, political, and values). In this section, we introduce two ideas: first, the 200-year present, and second, backcasting.

200-year present Elise Bounding introduced the idea of the 200-year present, in which one goes 100 years into the past and 100 years into the future (Boulding, 1988). The 200-year present is a time window that includes the year of the birth of people who turn one hundred today and the year when people born today turn one hundred years old. The two-hundred-year window allows one to grasp continuities and discontinuities that bridge how the past, present, and futures flow into each other.

The connection between Social, Technological, Economic, Environmental, Political forces and Values as they relate to futures can be challenging to imagine for many learners. One way to make STEEP+V forces more tangible to learners is to look at how STEEP+V forces historically played out in the past and to notice how current STEEP+V forces operate in the present times. Once one maps out the past and present STEEP+V forces, imagining how such forces might play out in alternative futures is easier.

Mapping STEEP+V forces over time allows one to notice how such forces played out in the past and are playing out in the present. One can then speculate on how such forces might play out in alternative futures.

Backcasting is a way to work from a desirable future scenario to a present situation. Backcasting begins with first establishing a desirable future. For a desired future, there should be clear goals and measurable indicators to assess progress toward each goal. Each benchmark goal in the future should have a pathway from the present. Along each path to describe the journey from the present to the future benchmark goal, there should be milestones and possible barriers over time. For example, the World Business Council of Sustainable Development (WBCSD) made a backcasting plan for a sustainable 2050. Figure 11.2 is a schematic illustration of the WBCSD plan showing the ten pathways, ten-year milestones, and possible barriers. The complete map is visible online *here* (WBCSD, 2009).

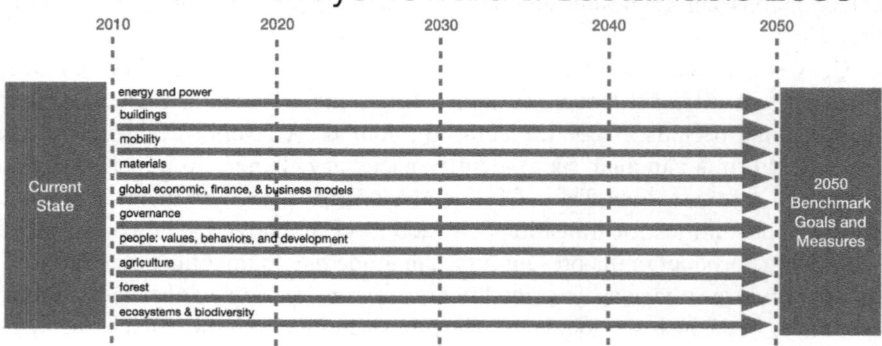

FIGURE 11.2 A schematic representation of the World Business Council on Sustainable Development (WBCSD) plan for a sustainable future by the year 2050. Nine pathways with benchmark goals and measures for the year 2050 and milestone goals by decade and possible barriers. (redrawn image credit © LELab.)

EXPLORING INTERNAL AND EXTERNAL FUTURES

So far, we've described futures out there, such as the Sustainable futures for the year 2050 from WBCSD, the ARUP global alternative futures scenarios for the year 2050, and so forth. How might one explore interior, individual, and collective futures? Ken Wilber developed a framework called the AQAL (All Quadrants, All Levels) Theory. AQAL is a framework for understanding individual and collective relationships. In Futures Studies, the Integral Theory perspective is called Integral Futures (Slaughter, 1998; e.g., Collins & Hines, 2010).

AQAL has two axes: on the horizontal axis, there are interior and exterior, and on the vertical axis, there are individual and group. The four quadrants describe individual, group, and interior and exterior views (Figure 11.3).

First, the "Individual Interior" is the "I quadrant" or the intentional subjective in the upper left quadrant. This "individual interior" world includes self-consciousness, values, purpose, and one's calling. Concerns include motivation, changes in people's values, perceptions, goals, etc.

Second, the "Individual Exterior," the "IT quadrant" or behavioral objective, and the individual exterior world are in the upper right quadrant. The IT perspective includes objective science, biology, behaviorism, and, more generally, "the facts." Concerns are observable changes in how people act externally, such as voting patterns, consumer behavior, etc.

Third, the lower right quadrant is the Group Exterior "ITS" and includes social systems and institutions and an inter-objective perspective (e.g., systems theory, deep ecology). ITS is the physical world, the world of systems and infrastructure.

Fourth, on the lower left quadrant, is the Group Interior. The "WE quadrant" is often called the cultural intersubjective and is the interior world of groups expressed

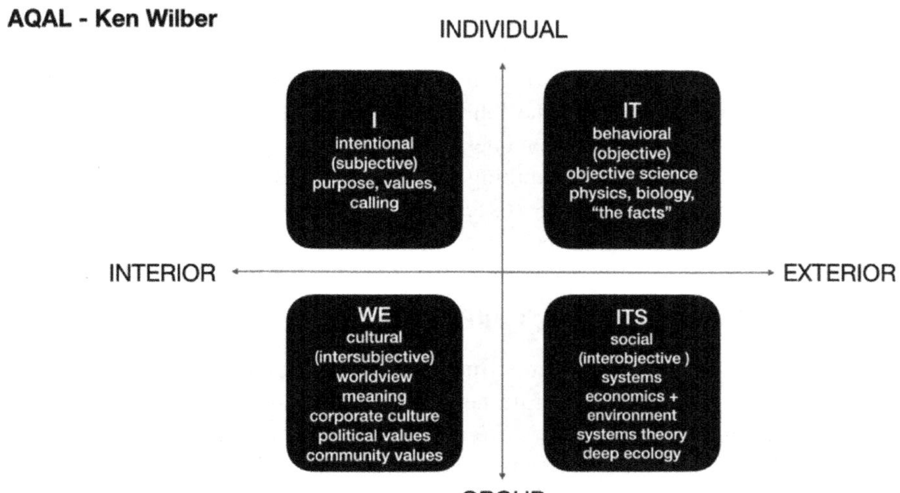

FIGURE 11.3 All Quadrants, All Levels (AQAL) framework by Ken Wilber. The horizontal axis is Interior – Exterior, and the vertical axis is Individual – Group. The four resulting quadrants are top left, Individual Interior, "I"; top right, Individual Exterior, "IT"; Bottom Left, Interior Group, "WE"; and bottom right, "ITS." (Image adapted from Esbjrn-Hargens, 2009.)

in culture. The WE quadrant includes culture and worldview, meaning, morality and values, and inter-subjective perspective (e.g., corporate culture, political values, community values).

Futurist Richard Slaughter saw Integral Theory as an opportunity to overcome flatland futures that only focus on limited aspects such as technological futures and limited perspectives (Slaughter, 1998). One of the critical ideas of Integral Theory in Integral Futures is integrating subjective experiences with the objective, inter-subjective, and inter-objective perspectives (e.g., Collins & Hines, 2010). Futurists explored three main ways to connect Integral Theory to futures studies theory: first, to shift perspectives in futures. Second, to apply Integral Theory into futures practice, and third, to intensely debate the role of Integral Theory and Integral Futures in futures studies (Collins & Hines, 2010).

I've found the premise of Integral Theory and Integral Futures helpful from a design perspective to imagine major transitions away from a current paradigm into new paradigms for issues of societal concern in the twenty-first century. Some examples include imagining the lifelong transitions of people towards zero-carbon lifestyles.

The design space for such societal transitions strategically engages all four quadrants and levels. Another example might be helping organizations of higher learning to embark on zero-carbon journeys with students and operations managers to invent roadmaps and pilot tests towards campuses with future zero-carbon operations. The four quadrants of AQAL help designers to imagine how the different parts of the whole might fit together and to seek strategic design actions. Personal futures exercises can help students explore their purpose, values, and calling through such futuring exercises. One can experience the other three quadrants and perspectives through diverse Design Futures exercises described in the previous sections. Teaching students to integrate the four quadrants and levels in their design practice remains challenging. An open question I am exploring is ontological in nature: How might designers go from thinking about futures abstractly, to embody Design Futures through action? In the next section, I explain how one might make Futures Scenarios more tangible and experienceable.

STRUCTURING DESIGN FUTURES EXPERIENCES

So far, I've described how particular futures thinking methods can be injected into design thinking processes to create new methods called *Dexign Futures.* Arnold Wasserman developed the term *Dexign.* The "X" replaces the "S" to signify an experimental design form. *Dexign* is an emergent form as designers take on societal challenges that require merging design thinking and futures thinking to align short-term design action with long-term sustainable vision goals. So far, in this chapter, I describe how to embed futures thinking methods into design thinking methods. Next, I explain how to render images of futures tangible and experienceable.

How might people experience, in tangible and meaningful ways, future worlds, scenarios, situations, futures artifacts, and futures stories? Established practices that make "Design Futures" tangible and experienceable range from Critical Design (Dunne, 2005; Dunne & Raby, 2001), Design Fiction (Bleecker et al., 2022; Sterling, 2005), Speculative Design (Dunne & Raby, 2013), Discursive Design (Tharp & Tharp, 2019), and Experiential Futures (Candy, 2010).

Such Design Futures traditions allow people to experience in a tangible way futures scenarios and futures ideas that would be otherwise ungraspable. In these cases, often the power of design is harnessed to make ideas about possible futures experienceable by prototyping fictional products, tangible futures experiences, and other forms of experiences linked to communications from futures. These uses of design's prototyping power to render futures experienceable are common.

More recently, two notable European research projects combining design and futures include Speculative Edu, exploring the limitations of speculative and critical design (Mitrović et al., 2021), and Fuel4design exploring design augmentation through futures methods (Morrison et al., 2021).

SUMMARY

There are plenty futures for Design Futures in HCI. Just as HCI is an expanding field, so are the emergent practices of Design Futures. I've described how HCI is moving forward in four waves or paradigms. More waves of HCI are likely in the futures as new technologies, societal needs, and planet-based realities emerge. I've described seven future thinking methods useful for designers interested in addressing societal-level challenges that play out through everyday design action: (a) explore drivers of change, (b) critique images of futures, (c) design for alternative futures, (d) deepen understandings with CLA, (e) mapping pathways linking past, present, and futures,

(f) exploring internal and external futures, and (g) structuring Design Futures experiences. I've provided examples of combining design thinking and futures thinking in seven sections. I hope this chapter will inspire other designers to find new ways to apply futures thinking in combination with design thinking, thus creating new tools and processes to engage with the challenges of the 21st century. While nobody can predict the future, exploring alternative futures helps one to prepare for what might happen.

REFERENCES

Bell, W. (2003). *Foundations of futures studies* (Vol. 1, rev. ed.). Transaction.

Bleecker, J., Foster, N., Girardin, F., Nova, N., Frey, C., & Pittman, P. & Near Future Laboratory (Organization). (2022). *The manual of design fiction*. Near Future Laboratory.

Bødker, S. (2006). When second wave HCI meets third wave challenges. In Proceedings NordiCHI 2006, pp. 14–18. ACM, New York. https://doi.org/10.1145/1182475.1182476

Bødker, S. (2015). Third-wave HCI, 10 years later—Participation and sharing. *Interactions*, 22(5), 24–31.

Boulding, E. (1988). *Building a global civic culture: Education for an interdependent world*. Syracuse University Press.

Candy, S. (2010). *The futures of everyday life: Politics and the design of experiential scenarios*. Doctoral Dissertation. University of Hawaii.

Cascio, J. (2012). Ten Rules for Creating Awful Scenarios. http://www.openthefuture.com/2012/08/ten_rules_for_creating_awful_s.html

Cascio, J. (2013). Bad Futurism. http://www.openthefuture.com/2013/02/humanity_plus_talk_bad_futuris.html

Collins, T., & Hines, A. (2010). The evolution of integral futures a status update. *World Futures Review*, 5–16. https://doi.org/10.1177/194675671000200303

Dator, J. (2009). Alternative futures at the Manoa School. *Journal of Futures Studies*, 14(2), 1–18.

Dator, J. (2019). What futures studies is, and is not. In *Jim Dator: A Noticer in Time*. Anticipation Science, vol 5. Springer, Cham. https://doi.org/10.1007/978-3-030-17387-6_1

Dunne, A. (2005). *Hertzian tales: electronic products, aesthetic experience, and critical design*. MIT Press.

Dunne, A., & Raby, F. (2001). *Design noir: the secret life of electronic objects*. August; Birkhäuser.

Dunne, A., & Raby, F. (2013). *Speculative everything: Design, fiction, and social dreaming*. MIT Press.

Esbjrn-Hargens, S. (2009). "An Overview of Integral Theory: An All-Inclusive Framework for the 21st Century," Integral Institute, Resource Paper No. 1 (March 2009): 1.

Fry, T. (1999). *A new design philosophy : an introduction to defuturing*. UNSW Press.

Fry, T. (2008). *Design futuring : sustainability, ethics and new practice*. Berg.

Gidley, J. M. (2017). *The future: A very short introduction*. Oxford University Press.

Gordon, A. (2009). *Future savvy: identifying trends to make better decisions manage uncertainty and profit from change*. American Management Association.

Harrison, S., Tatar, D., & Sengers, P. (2007, April). The three paradigms of HCI. In Alt. Chi. Session at the SIGCHI Conference on human factors in computing systems San Jose, California, USA (pp. 1–18).

Inayatullah, S. (1998). Causal layered analysis: Poststructuralism as method. *Futures*, 30(8), 815–829.

Intergovernmental Panel Climate Change (IPCC). (2023). https://www.ipcc.ch/reports/

Lopes, A. G. (2022). HCI Four Waves Within Different Interaction Design Examples. In: Bhutkar, G., et al. Human Work Interaction Design. Artificial Intelligence and Designing for a Positive Work Experience in a Low Desire Society. HWID 2021. IFIP Advances in Information and Communication Technology, vol 609. Springer, Cham. https://doi.org/10.1007/978-3-031-02904-2_4

Mitrović, I., Auger, J., Hanna, J., & Helgason, I. (2021). Beyond Speculative Design: Past – Present – Future, SpeculativeEdu; Arts Academy, University of Split.

Morrison, A., Celi, M., & Cleriès, L. (2021). Anticipatory design and futures literacies. In Proceedings of CUMULUS ROME 2020. https://cumulusroma2020.org/

Rijn, M. van, & Burgt, R. v. d.. (2021). *Explore the big picture: Forces shaping the future of humanity*. Pearson.

Schemel, S., Simunich, J., Luebkeman, C., Ozinsky, A., McCullough, R., & Bushnell, L. (2019). *Four plausible futures 2050 scenarios* (p. 68). ARUP. https://www.arup.com/perspectives/publications/research/section/2050-scenarios-four-plausible-futures

Schwartz, P. (1991). The art of the long view (First). Doubleday/Currency.

Scupelli, P. (2021). Teaching designers to anticipate future challenges with causal layered analysis. IASDR 2021: With Design: Reinventing Design Modes. December 5–9, 2021. https://www.sd.polyu.edu.hk/iasdr2021/abstracts/download.php

Scupelli, P. (2022). Does when and how design students learn causal layered analysis matter? *Journal of Futures Studies*, 27(2), 28–41.

Scupelli, P. (2023) Teaching to transfer causal layered analysis from futures thinking to design thinking, in De Sainz Molestina, D., Galluzzo, L., Rizzo, F., Spallazzo, D. (Eds.), IASDR 2023: Life-Changing Design, 9–13 October, Milan, Italy. https://doi.org/10.21606/iasdr.2023.383

Slaughter, R. A. (1998). Transcending flatland: Implications of Ken Wilber's meta-narrative for futures studies. *Futures*, 519–533. https://doi.org/10.1016/S0016-3287(98)00056-1

Sterling, B. (2005). *Shaping things*. MIT Press.

Tharp, B. M., & Tharp, S. M. (2019). *Discursive design: Critical, speculative, and alternative things*. MIT Press.

World Business Council for Sustainable Development (WBCSD) (2009). Nine Pathways to a Sustainable 2050. https://www.wbcsd.org/vision-2050/

12 Ecological Thinking in HCI

Xiaojuan Ma

INTRODUCTION

> "The First Law of Ecology: Everything is Connected to Everything Else" – by Barry Commoner in the best-selling book "The Closing Circle: Nature, Man, and Technology"
>
> **(Commoner, 2020)**

The term "ecology" was originated from Greek as a combination of "oikos" (meaning household) and "logy" (i.e., a common ending derived from logos to suggest a systematic discourse or study), according to Raymond Williams, a cultural theorist (Williams, 1976). In his book "Keywords: A vocabulary of culture and society," Williams defined ecology as "the study of the relations of plants and animals with each other and with their habitat...a characteristic living place." The scientific use of the word ecology can be dated back to the 1860s, when the German zoologist Ernst Haeckel established a new discipline of ökologie, expanding it from "the study of households" to "the science of the relations of the organism to the environment, including the study of all the conditions of existence, both organic and inorganic" (Kingsland, 1995). The term did not enter everyday English until mid-20th century, as evidenced by Linda Lear in her biography of the biologist Rachel Carson (Lear, 1998). Since the 1960s, Williams observed an extension of the notion of ecology to economics, politics, and social theory being "reinterpreted from a central concern with human relations to the physical world as the necessary basis for social and economic policy" (Williams, 1976). To date, ecology can be broadly defined as a study that aims to characterize the physical (natural) and social aspects of habitats where humans endeavor to live well together (Conley, 2006), understand factors that facilitate or hinder such living (Code, 2006), dissect reasons why people strive or fail to realize this vision, and propose possible actions.

The notion of ecology has a long history of being employed in human-computer interaction (HCI) research (Blevis et al., 2015). It is applied at different levels and consequently leads to different working definitions and practices of ecological thinking when computing systems are introduced into the human world.

The first line of HCI research adopts concepts and principles from ecological psychology (Reed, 1996) at a micro-scale to inform the design of specific interfaces and interactions (Kaptelinin, 2014; Norman, 2004). Researchers postulate that humans learn to interact with a piece of technology through the continuity of perception and action, which is an embodied, situated, and non-representational experience similar to how an actor, human or animal, discovers the actionable properties of the world (Norman, 1988, 1990).

DOI: 10.1201/9781032693606-16

The second line of HCI research approaches ecological thinking at a meso-scale, using the concept to describe and analyze a collection of entities (e.g., products, media, or technologies) of interest. Researchers consider these collective entities and their contexts as an "ecosystem," a "habitat," or a "milieu" – "the site, habitat, or medium of ecological interaction and encounter" (Hayden, 1997). They develop frameworks such as "information ecologies" (Nardi & O'Day, 2000), "artifact ecologies" (Bødker & Klokmose, 2011), "media ecologies" (Fuller, 2005), and "product ecology" (Forlizzi, 2008). These frameworks allow HCI scholars to investigate the complex relationship between an individual technology and other designs or the complex configuration of a system composed of multiple designs (Blevis et al., 2015), by analyzing "the compositions of relations or capacities between different things" and "the capacities for affecting and being affected that characterize each thing" as in Deleuze's naturalism (Deleuze, 1988).

The third line of HCI research maintains the common definition of ecology as the study of "interrelationships of living things and their environments" (Hayden, 1997). Researchers are interested in "of which concepts, practices, and values best promote the collective life and interests of the diverse modes of existence inhabiting the planet" (Hayden, 1997). In other words, they are interested in understanding how the emergence of technologies create, reshape, or destruct the Earth's combined natural-social habitats and the underpinning ethical and political issues. This body of works focus on sustainability issues in the lifecycle of designing, developing, deploying, and disposing of technologies (Hazas & Nathan, 2017), trying to optimize humans' living conditions while avoiding ecological disasters including but not limited to pollution, extinction of living species, climate disturbance, pandemic, massive unemployment, escalation of racism, and wars.

Regardless of how ecological thinking is practiced in HCI research and which ecological theory is referenced, the "habitat" of study, whether it refers to an individual design, a collective system, or a living environment, "joins up not only its spatiotemporal but its qualitative planes or sections" (Deleuze & Guattari, 1994). Many ecological HCI studies thus take a special interest in one or more of the dimensions of space, time, and entity (artifact or nonhuman organism), as well as their interplay. In the rest of this chapter, we discuss how each of the dimensions can play a role in the study of computing technologies for human use, as an analytical lens or a design material.

SPACE

> Place and the scale of space must be measured against our bodies and their capabilities.
>
> **– By Gary Snyder**

All aspects of an ecological system unfold in space. The dynamic eco-processes happen over space and time. While we say all discrete entities, living or nonliving, human or nonhuman, share the world on Earth, their immediate neighborhood has the strongest impact on them as they tend to interact with other entities within a certain spatial range. The field of spatial ecology, first coined by Tilman and Karieva (Tilman & Kareiva, 1997), refers to "the study and modeling of the role(s) of space on ecological processes … that in turn affects ecological patterns" (Fletcher et al., 2018). Subdisciplines under spatial ecology are looking at the possible influences that factors such as spatial scales, movement in space, and spatial heterogeneity may directly or indirectly have on the functioning and outcomes of ecosystems. They are particularly interested in how endogenous and exogenous processes play out and "result in the spatial patterns observed at different levels of organization though space" (Fletcher et al., 2018). Endogenous processes refer to "the dynamics of each ecological entity … and the interactions among entities within and across species," whereas exogenous processes concern "the response of organisms to environmental factors that are themselves spatially structured" (Fletcher et al., 2018).

Human activities take place in a multitude of spaces. People are inclined to depict our relationship with spaces in reference to our body and capabilities, as noted in the quote of Gary Synder. For instance, Barbara Tversky classified people's mental representations of the perceived spaces as "the space of the body," "the space around the body," "the space of navigation," and "the space of graphics" (Tversky, 2003). This classification is based on the functions (together with the corresponding things and their spatial relations) served by the different spaces. Daniel R. Montello proposed to use scale, "ratio between the dimensions of a representation and those of the thing that it represents," to categorize psychological space into "figural" (i.e., smaller than the body and apprehensible without locomotion), "vista" (i.e., as large or larger than the body and apprehensible without locomotion), "environmental" (i.e., larger than the body but only apprehensible with locomotion), and "geographical" (i.e., larger than the body and not directly apprehensible through locomotion) spaces (Montello, 1993). HCI researchers further extended this categorization to include a fifth class of spatial scale "panoramic" (i.e., larger than the body, apprehensible without locomotion, but not visible in their entirety from any orientation) and made a "vista" – the always visible part – a subspace of it (Barba & MacIntyre, 2011). The following case study is a preliminary investigation of how the different spatial scales and relationships affect the endogenous process of interacting with a computing system.

Case Study on Scale. When people interact with conventional screen-based computing devices, they are likely to operate in the figural space (e.g., checking the phone) and the vista space (e.g., reading on a wall display) in front of them. The advancing augmented reality (AR) technologies empower situated, interactive visual analytics beyond the 2D screen, making full use of the multitudes of spatial scales (Figure 12.1 (A)). We conducted an exploratory study on the way users perform logical reasoning tasks on AR visualizations displayed in two different scales commonly found in daily activities (i.e., the size of a room versus the size of a table, see Figure 12.1 (C)) (Sun et al., 2018). Modeling after classic logical reasoning puzzles, we asked the participants to infer the seating arrangement of eight characters based on a set of seven clues (e.g., person G sits in one of the corners; F and H come from the same group, and they take the same course(s)). Some of the clues are visualized in a node-link graph overlaid on top of the actual seats (room-scale) or a floor map of the room (table-scale) as pictured in Figure 12.1 (B). We captured how users position

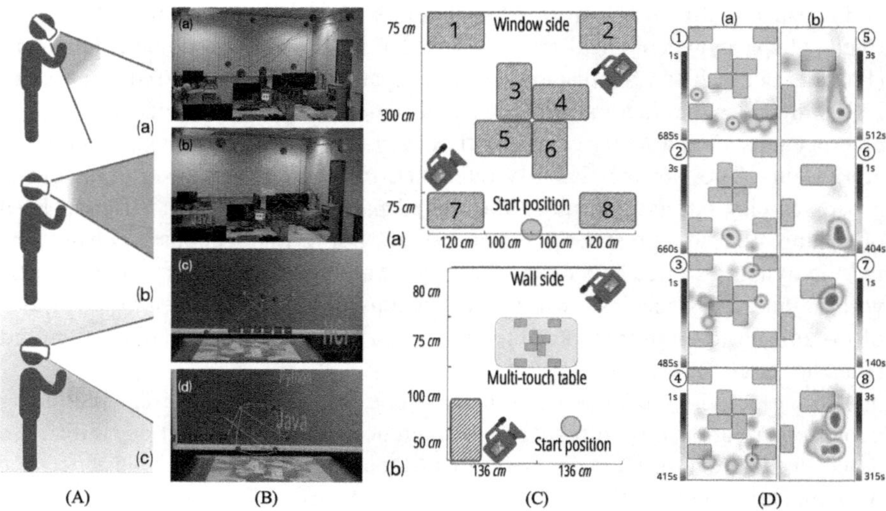

FIGURE 12.1 (A) The schematic diagram for three spatial relationships: (a) figural, (b) vista, and (c) panoramic. (B) Screenshots of the virtual visualization model augmented on physical objects during exploration in the AR context: (a) and (b) are the room-scale and (c) and (d) are the table-scale. (C) Seating arrangement and experimental setup under (a) room-scale and (b) table-scale. (D) Heatmap of the duration spent in the room-scale (column (a)) and table-scale (column (b)) based on participants' movement trajectories captured in video recordings. Plots are arranged by the primary spatial relationship adopted, i.e., figural space: ①⑤; vista space: ②⑥; panoramic space: ③⑦; and mixed space: ④⑧ (Sun et al., 2018).

themselves in respect to the situated visualization and dynamically move in space during interactive analytics (Figure 12.1 (D)). We further analyzed the possible reasons behind users' locomotion and the issues they encountered. We observed that people adjusted the size of the visual stimulus or changed their own proxemics and perspectives in space for better manipulation in a certain spatial scale based on the task to complete. Some liked to stick to one scale throughout the process while others switched between various spatial arrangements to take advantage of each configuration. Different spatial relationships with the visual stimulus may influence people's analytical mindsets and tactics.

The scenario rendered in this case study can be considered as a kind of reality-based interaction (RBI) with post-WIMP ("window, icon, menu, pointing device") interfaces (Jacob et al., 2008). RBIs "draw strength by building on users' pre-existing knowledge of the everyday, non-digital world to a much greater extent than before… [and] employ themes of reality such as users' understanding of naïve physics, their own bodies, the surrounding environment, and other people" (Jacob et al., 2008). The above case study was conducted in a rather private office space with no other people around. The following case study looked into the complex situations when situated visual analytical activities occur in public spaces shared by many people. As illustrated in Figure 12.2 (A), if the first case study fell into the third level of RBI, the second study reached the fourth level of social awareness and looked at

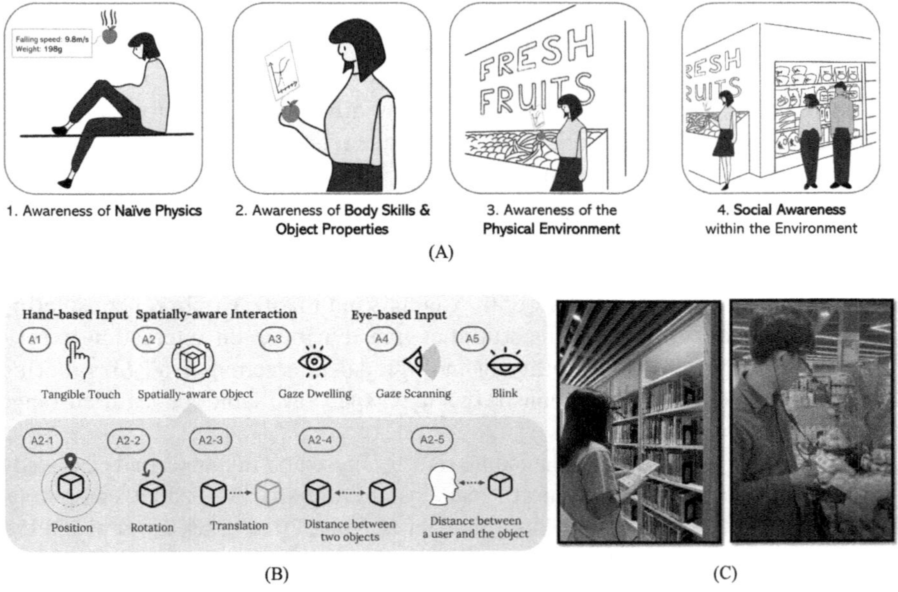

FIGURE 12.2 (A) The four levels of awareness during interactions with situated visualization from users' perspective, derived from the unified reality-based interaction (RBI) framework. (B) Possible interaction input modalities for selected public scenarios shown in (C), namely library and market (Zhu et al., 2024).

the exogenous processes – "the response … to environmental factors that are … spatially structured" (Fletcher et al., 2018).

Case Study on Social Acceptance in Public Space. The release of Apple Vision Pro stimulates creative usage of "spatial computers" in a wide range of settings, private or public. There have been videos of people playing with Vision Pro on the street, in coffee shops, on the planes, etc. This illustrates a possible future that situated visualization can support people's routine activities. Our recent project "Make Interaction Situated" targets such a scenario. "As users' awareness and understanding of the surrounding context gradually broaden, their adoption of the interactions with situated visualization can be more aligned with the real-world environment" (Zhu et al., 2024). This alignment is not just about matching the interactions with the action abilities that the environment can provide, but also gaining social acceptance (a.k.a. social acceptability, i.e., "deemed to be appropriate, by both the user and any observers, in the context in which they are carried out") (Montero et al., 2010). It is because interactions in public settings does not only occur between users and intended physical entities but also within the entire environment (Rico & Brewster, 2010), and thus the design of interactions with any situated computing devices should ensure that users are willing to engage with them with the presence of other people in the same space. We conducted a formative study to extract users' contextual needs of situated visualization in terms of data, display, representation, and interaction in two common public spaces (i.e., a library and a grocery store, see Figure 12.2 (C)). We then compiled a list of potential user input modalities that are publicly acceptable and technically feasible for a spatial computer such as Apple Vision Pro. These modalities include hand-based and eye-based interactions, as well as spatially aware interactions leveraging information about distance, orientation, movement, and position (Figure 12.2 (B)). We carried out an iterative design process to select the most suitable interaction modalities according to the data description and the analytical operation to perform. Our field evaluation with a research prototype suggested that the proposed interactions were publicly acceptable, flexible, and practical in the real world for the participants. We obtained several insights from this research: (1) considering the impact of environmental uncertainties and dynamics on interaction experiences, (2) exploiting the flexibility of multimodal interactions, and (3) tailoring interactions to environmental constraints (e.g., space available for required range of motion) and user acceptance.

The interactive activities in the public settings presented in our second case study happened primarily in the figural, vista, panoramic spaces. Spatial computing, another HCI interaction framework (Shekhar & Vold, 2020; Shekhar et al., 2015), further extends the spatial confinement of interactions to the environmental and even geographical spaces. Through the use of positioning technologies and digital navigation systems (e.g., GPS and Google maps), humans' understanding of and navigation through locations can reach a new level, enabling location-based interactive experiences at a much bigger spatial scale.

TIME

It is common to take an evolutionary approach to ecological analysis, studying the "patterns in nature, … how these patterns came to be, how they change … in time, why some are more fragile than others" (Kingsland, 1995). To ecologists, "the full diversity of life is exhibited through natural processes of change and becoming" manifested over time, and "a complete milieu … possesses a relative rather than absolute equilibrium" (Hayden, 1997).

Time as an Analytical Facet

Similarly, the field of HCI acknowledges the dynamic life cycle of technologies and several HCI scholars advocate for the use of time as a fundamental analytical facet of the design and application of computing systems (Odom et al., 2018a; Rahm-Skågeby & Rahm, 2022). As Mazé and Redström summarized, HCI researchers and designers need to "investigate what it means to design a relationship with a computational thing that will last and develop over time" whose form "is fundamentally constituted by its temporal manifestation" (Mazé & Redström, 2005).

One typical way for HCI researchers to observe and measure the "change and becoming" – ecological concepts – of technologies in human context across different time points is through longitudinal studies that last days, weeks, months, or even over a year (Kjærup et al., 2021; Vaughan et al., 2008). Kjærup and colleagues surveyed the longitudinal studies reported in publications at the premier international HCI conference The ACM CHI Conference on Human Factors in Computing Systems (CHI) between 1982 and 2019 (Kjærup et al., 2021). Despite that most HCI research focuses primarily on short-term interactions, they identified 106 CHI papers for analysis out of 38 years of conference proceedings. The themes of these longitudinal studies are (in descending order of the number of related publications): understanding users, interface artifacts or techniques, systems or tools, methodology, and theory. Analysis revealed that 66% of the surveyed papers explicitly reported results on change (e.g., evolution of user behavior or the design of inquiry) or stability (e.g., plateauing in performance) over time (Kjærup et al., 2021). Researchers applied quantitative and qualitative methods to capture the temporal characteristics of the use (and non-use) of computing systems, for instance, tend and outlines of time-dependent variables, emergence or disappearance of practices and effects before, during, and after long-term usage, cause and effect unfolded on a time scale, and historical dynamics and shifts, to name a few. These are examples of exercising ecological thinking at the micro- and meso-levels.

Still, the typical study duration of longitudinal research is short compared to the lifespan of human beings, not to mention the history of planet Earth. The notion of "deep time" is thus introduced from environmental archeology (Heringman, 2015; Irvine, 2014) to HCI to draw research attention to the long-term interactions at a geological timescale. Deep time "refers to temporalities that include the fundamentally perdurable geological processes of the Earth – effectively considering the pace, rhythm, causalities, and materialities by which durable ecological changes occur" (Rahm-Skågeby & Rahm, 2022). This concept stresses the fact that the flourishing of human societies and the evolution of technologies depend on many natural resources that "have formed over geological time spans under conditions that are difficult or impossible to reproduce" (Cervato & Frodeman, 2012). It is thus necessary to situate the understanding of HCI in Earth history by promoting the deep time design thinking. This means "exploring the interactions between humans, technologies and geological temporalities" and "reflecting on the long-term impacts of computer technologies" at different temporal resolutions ranging from daily life to geological timescales (Rahm-Skågeby & Rahm, 2022). It is a kind of meso-to-macro-level ecological thinking that allows HCI scholars to identify and inspect Anthropocene epochs that are significantly shaped by human activities involving technologies, understand how and to what extent the design and use of computing systems may contribute to the contemporary ecological development and crisis, and inform social, cultural, ethical, economical, and political decisions.

Time as an Experience to Be Designed

Temporality, "the state of existing within time" (Odom et al., 2018a), and the perception of time are a critical component of user experiences. HCI has long been considering time-related design objectives, dealing with temporal constraints, and exploring the possible impact on the perception of time.

Fast and Slow

> *"从前的日色变得慢*
> *车，马，邮件都慢*
> *一生只够爱一个人*
> *Days were slower in the past.*
> *Carriage, horse, and mail did not reach fast.*
> *You need your lifetime to just love the person who is right."*
> *– By Mu Xin in "《从前慢》Slower Days in the Past"*

This poem from Mu Xin illustrates the change of pace in our lives. Efficiency and productivity have become the common design goals in modern human societies. Many technologies feature fast interactions that can be quickly learned and applied and use measures such as task completion time, word per minute, throughput, speed, and latency to optimize performances.

Case Study on Fast Technology. To demonstrate the ecological thinking around the design of technologies that expect speedy interactions, we present a case study on the effect of latency, "the time interval between a user query and the system's response" (Wu et al., 2019), on user experiences. The rapidly advancing speech and language technologies (e.g., large language models and generative AI) enable human users to interact with computing systems through voice-based conversations. Many companies provide cloud APIs to a large model that a system can call to obtain high-quality processing and synthesis of speech, which however may suffer from network latency. Alternatively, one can employ a smaller local model to ensure the accessing speed with sacrifice in output quality. The latter solution is more economically sustainable for small business and communities with ecological limits (Knowles et al., 2018).

Given that voice interaction demands high responsiveness, we conducted a study to investigate the trade-off between a robot's quality of voice and response time and its impact on user experience (Peng et al., 2020). Figure 12.3 shows the robot we used, the pipeline of processing user input and generate corresponding output speech, and two versions of implementation (i.e., bot1 using online services that produce higher-quality voice at a longer delay versus bot2 using offline models that are faster but of a lower quality). Experimental results suggest that users' satisfaction with the conversations may maintain up to four seconds of delay and are likely to drop considerably after that. One interesting finding is that a higher-quality voice may even lead users to feel that the robot responds faster. If users have experienced a high-quality voice previously, reducing the delay at the cost of reducing voice quality could not improve user experience. This case study reveals that temporal and qualitative aspects of an interaction are interrelated and the order of exposure to different conditions across the course of interaction may impact their effectiveness on enhancing user experiences during the prolonged or repetitive use of the service.

Conversational agent is not the only piece of technology that cares about latency. Extended reality (XR) applications also demand low latency to boost the perceived illusion of stability, reduce the stiffness of virtual objects and the risk of motion sickness, and ensure users' sense of presence in immersive environments (Wu et al., 2019).

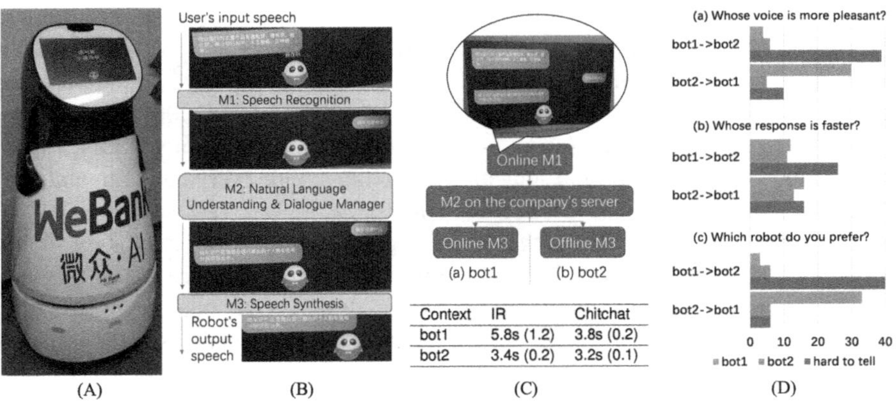

Context	IR	Chitchat
bot1	5.8s (1.2)	3.8s (0.2)
bot2	3.4s (0.2)	3.2s (0.1)

FIGURE 12.3 (A) The robot "Xiaowei" used in the study. (B) The computational models of robot Xiaowei's dialogue system and the two methods of implementing a speech synthesis model (M3) in the robot: bot1 invokes M3 online, which has a higher-quality voice but is slower; bot2 invokes an offline M3, which is faster but has a lower-quality voice. (C) The average (SD) delay of responses (in seconds) in the three-round conversations with bot1 and bot2 in the contexts of information retrieval (IR) conversation and chit-chat. (D) User preference for bot1 and bot2 when they interact with the robots in different orders (Peng et al., 2020).

Sometimes latency is unavoidable due to technical constraints, designers hence need to introduce smart designs to hide the long wait time; for instance, a voice agent can insert backchannel cues to signal active listening and continuation before the return of the actual response.

On the other hand, there are occasions where HCI researchers and designers deliberately introduce delay into the interaction. For instance, the average delay of fingerprint authentication on smartphones was around 700 ms in 2020 (Wu et al., 2020). The processing speed on this task already surpasses that of humans and thus some users, especially older adults, have doubts about the reliability of the technology – whether it is actually doing the job. In such situations, an artificial delay can be added to make the process more visible to users. In another example, "slowing down the process" as one of the cognitive forcing functions is employed in AI-assisted decision-making (Buçinca et al., 2021). HCI studies show that simply delaying the presentation of AI recommendations can stimulate humans' system 2 thinking (Daniel, 2017) to decrease their overreliance on AI and boost decision accuracy (Buçinca et al., 2021; Park et al., 2019).

The aforementioned works tailor their designs to fit the temporal process of human perception and cognition. The slowness that is involved is only in a period of seconds or minutes and the ultimate goal is still productivity. Besides tools to be applied in specific situations for a limited period to take away time required on a job, HCI researchers also highlight the need for "creating technology that surrounds us and therefore is part of our activities for long periods of time" (Hallnäs & Redström, 2001). This call results in a design agenda of slow technology that aims at "reflection and moments of mental rest rather than efficiency in performance" (Hallnäs & Redström, 2001). A slow technology differs from a fast one in that it opens people up for time presence by exploiting "slowness in learning, understanding and presence to give people time to think and reflect" (Hallnäs & Redström, 2001). In this process, slowness has also been shown to foster anticipation (Odom et al., 2014), social connection (Odom, 2015), and longer-term relations with everyday computational objects (Odom et al., 2018b; Odom et al., 2019).

Case Study on Slow Technology. Users engaged with a slow technology should not feel that it is time-consuming or time-wasting; rather, they are meant to spend quality, meaningful time interacting with the design. One common method to achieve this effect is through (aesthetic) expression of functionalities, which intends to lead users to concentrate on the reflective aspects of the functionalities (Hallnäs & Redström, 2001). Along this line of thinking, the "Blossom" project proposed the idea of "embedding asynchronous voice message systems into specially designed tangible interfaces as a complement to the existing synchronous communication systems" to promote reflection and action on intergenerational communication (Zhao et al., 2016). In our research-through-design exploration, we learned that older adults who do not stay with their children and/or grandchildren would like to maintain a sense of being together through explicit communication. However, they may have reservations about the timing to call or text the younger generations, afraid of interrupting their busy schedule. At the same time, if the older adults do not receive a call or message back after some time, they are likely to feel being cut off and even forgotten. In view of the complex facets of the problem, rather than creating a fast technology that pushes immediate, synchronous communication between older adults and their children / grandchildren, we proposed to embed asynchronous, intergenerational interactions into a common object in the household – a pair of vases of metal flowers (see sketch illustration in Figure 12.4). One of the vases is placed in the older adult's home and the other is kept by the younger generation.

To be more specific, whenever the older adults want to leave a message to their children or grandchildren, they can roll the handle on their vase to record their voices and press the only button to send it. Upon receiving a new message, one of the flowers in the other vase will close up and turned into a bud (see Figure 12.4), indicating that an unread new message is stored. The state of the flower pedals serves as an ambient cue to remind the younger generation to take further action. Its look and its quiet existence in the environment mitigate the technological anxiety and the intended receiver does not need to listen to the voice message right away. They can choose to play it at a good time by clicking the button. The flower will blossom again after that (see Figure 12.4). To provoke reflection and reminiscence on the intergenerational relationship, we experimented with an asymmetric feedback mechanism on the family photo album attached to the vases. On the older adults' side, we design a positive reward scheme. Every time their intended recipient listens to their voice message, the family photo in the frame will become more visually vivid and colorful. The younger generations' side, on the contrary, adopted a negative feedback. If they neglect or refuse to respond to the messages left on the vase for a certain period of time, the figures of their parents / grandparents in the family photo will gradually fade out (Figure 12.4). This design resembles the

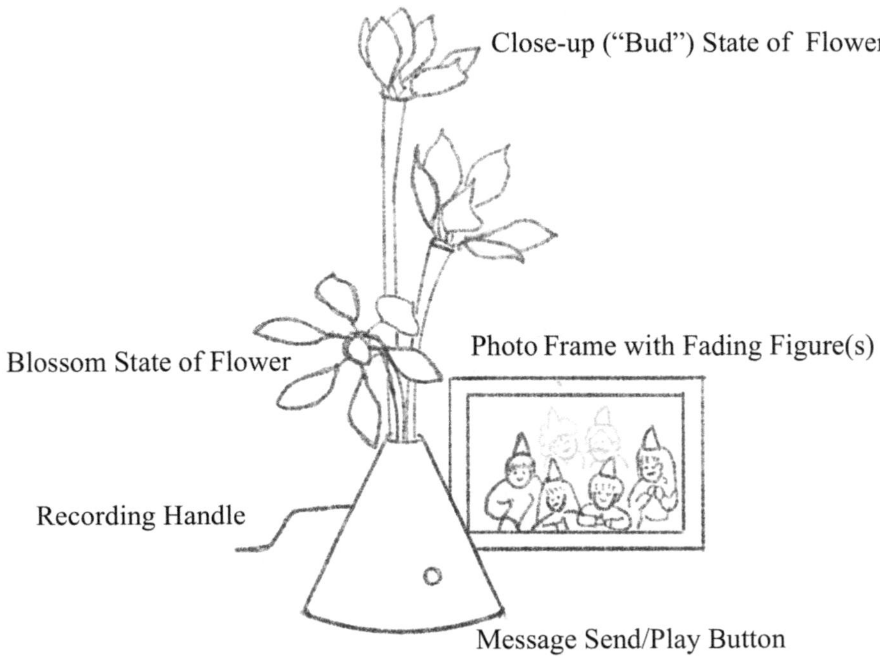

FIGURE 12.4 Blossom system. The sketch depicts one of the two vases with flowers and family photo that can fade out. It also illustrates both the close-up status of a flowers and the blossom status of a flower (Zhao et al., 2016).

disappearance of memory of time, expressing a feeling of sadness and guilt. With the slow technology of Blossom, its dynamic flowers, and fading-out/coloring of the family photo, we hope to inspire engagement in asynchronous, intergenerational communication through an ambient, subtle, but expressive design.

In brief, both fast and slow technologies can benefit users, though from different perspectives.

Past, Present, and Future

> Learn from yesterday, live for today, hope for tomorrow. The important thing is not to stop questioning.
>
> **– by Albert Einstein**

The perception of time and deep time naturally entail the concepts of past, present, and future. There have been many HCI works on the reservation of and reflection on the past, the appreciation of and immersion in the present, and the making of the future, as depicted in the above quote by Albert Einstein.

In their paper "This changes sustainable HCI," Knowles and colleagues summarized nine categories of sustainable HCI (SHCI) research when trying to compile a coherent research agenda for the community (Knowles et al., 2018). Among them, two categories of emergent SHCI studies have a future-looking element: "future proofing" and "inspiring with future visions" (Knowles et al., 2018). Research that is "future proofing" considers ecological limits, creates technologies that are less affected by resource collapse, and explores new self-sufficient computing paradigms and infrastructures. The goal is to enable human societies to "adapt to a future of scarcity in a comparatively just way" (Knowles et al., 2018). Works on "inspiring with future visions" take the approach of picturing a sustainable future to motivate necessary movements that can create a path to it. They commonly use methods such as design fictions in the process. Several chapters of this book have in-depth discussions about the topic of design futures. Hence, in this chapter, we focus more on ecological thinking in past-informed and present-focused designs.

Lorraine Code postulated that the notions of ethos and habitus "contribute to a conceptual frame for the modalities of ecological thinking and for shaping a conception of ecological subjectivity" (Code, 2006). In particular, habitus was defined by Pierre Bourdieu as "a product of history" that "produces individual and collective practices – more history – in accordance with the schemes generated by history" (Bourdieu, 1990). Conceptually, the habitus embraces "an infinite capacity for generating products – thoughts, perceptions, expressions and actions – whose limits are set by the historically and socially situated conditions of its production" (Bourdieu, 1990). To ecologists, many of the processes that they are keen to understand – geographical distribution, evolution, succession, etc. – are historical (Kingsland, 1995). They are using insights into the past to resolve emergent ecological issues.

In ecological thinking, there are different ways to view the relationships between past and present. From a sustainability perspective, "the conditions of the present are created by using what has been created over long periods of time. This 'loan' from history is then repaid in an altered state. The present (where time is money) is thereby robbing both the past and the future of its ecological sustainability" (Rahm-Skågeby & Rahm, 2022). Another angle is the "logic of practice," which conceives the present to be the internalized (embodied) history, "an active presence of past experiences, which, deposited in each organism in the form of schemes of perception, thought and action" (Bourdieu, 1990). The following case study can serve as a demo of how these views are weaved into a past-informed design.

Case Study on Past-informed Design. "A Postcard from your Food Journey in the Past" (Sun et al., 2020b) is a project that aims to facilitate in-depth, non-judgmental self-reflection on food intake and the associated emotional episodes in the past. It does not require users to effortfully log their meals every day. Instead, we propose to leverage people's food posts (i.e., food pictures and textual comments) on social media to derive information related to their nutrition values and emotional states. This is automatically done through computer vision, natural language processing, and data mining techniques. The output constitutes entries of personal data records which are then encoded into the form of a digital postcard. This is to mitigate the pitfall of existing statistical design employed by common self-tracking tools for dining experiences, avoiding evoking negative feelings, judgment, or obsession. The whole pipeline is shown in Figure 12.5 (A). In particular, we explore the use of associative and metaphorical visual encoding of the nutrition and emotion data. As illustrated

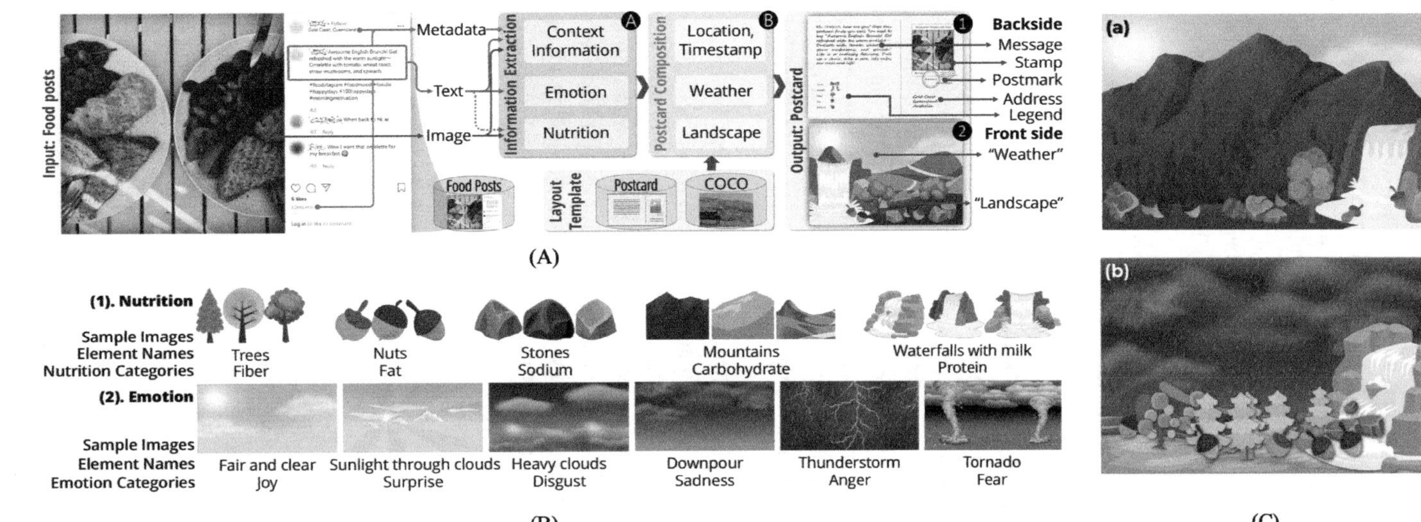

FIGURE 12.5 (A) The overview of the proposed pipeline to automatically generate postcards based on food posts data. (B) The encoding scheme and sample design materials applied in the front side of the postcard visual design. (C) The front side design of two representative postcard designs: (a) reveals a joyful status with high carbohydrate intake (Participant 1, male, 22); (b) indicates a sad status with a lot of fiber and protein in the diet (Participant 13, female, 22) (Sun et al., 2020b).

in Figure 12.5 (B), we apply five common elements found in a landscape to denote five key nutrients: trees-fiber, nuts-fact, stones-sodium, mountains-carbohydrate, and waterfalls with milk-protein. These mappings are intuitive to users as they are built upon the associations between the nutrients and their common sources in nature. Following real-world common sense, we adjust the number or size of the visual elements according to the corresponding data values and then arrange them into a landscape picture. Users' emotional state, classified into the six basic categories of emotion, is represented by the weather in the background: fair and clear sky-joy, sunlight through clouds-surprise, heavy clouds-disgust, downpour-sadness, thunderstorm-anger, and tornado-fear (see examples in Figure 12.5 (C)). The picture is painted on the front side of the postcard with the original food post, the visualization legend, and other meta information such as location and time of the post are printed on the back. The concept of a mailed postcard implies that this is information from the past, and the landscape encoding communicates the idea that the human body and the planet earth are both forms of milieu – a word in ecology that describes "all that is involved in the interactions between elements, compounds, energy sources, and organisms from the molecular to the molar levels" (Hayden, 1997). We collected data from 20 students who posted their detailed food-emotion information on social media over a course of three weeks and then showed them the auto-generated postcards for their interpretation and feedback. Results show that our design can indeed evoke curiosity in users and stimulate their reflection on data to facilitate recall of past practices, reflection on actions to trigger future reactions, and reflection on value about personal pursuit.

While the previous conceptualization situates the present in the past, there are scenarios where users are expected to accept and enjoy the present moment without contemplating the history or trying to reason about any complex relationships. A typical example of such scenarios is the practicing of mindfulness for stress reduction and wellbeing. HCI researchers proposed a 4-level framework of digital mindfulness – the use of gadgets and apps to support achieving the mindful state of mind (Zhu et al., 2017). At the first three levels, the digital tools are designed in a way that users can gain their presence through these instruments. It is postulated that at the highest level, a design for mindfulness should enable users "to be present without instrumental concerns," i.e., achieving a sense of presence-in and presence-with (Zhu et al., 2017). Zhu offered an analogy to digital mindfulness 4.0: "that of nature as a gateway to presence" (Zhu et al., 2017).

This analogy indicates another noteworthy contemporary ecological theory – deep ecology that was first proposed by the philosopher Arne Naess (Naess, 1990, 2017) and further developed by other scholars. Deep ecology, different from Deleuze's naturalism, views the world as an "unbroken wholeness" (Naess, 2017) "with no discontinuities or boundaries between human and nonhuman nature" (Hayden, 1997). This theory stresses the "intrinsic, spiritual identification of self and nature" (Hayden, 1997). This line of ecological thinking coincides with the traditional Chinese philosophy and aesthetics theory of "harmonious unity of Heaven and humanity (in Chinese: 天人合一)". Zhuang Zhou, an influential Chinese philosopher and the pivot figure of Daoism in the late 4th century BC, wrote in his book "Zhuangzi – The Adjustment of Controversies《庄子·齐物论》" that "Heaven, Earth, and I were produced together, and all things and I are one (in Chinese: 天地与我并生, 万物与我为一)." Although the analogy was drawn upon nature, Zhu argued that "not only nature offers possibilities for presence-in and presence-with but also many human aesthetic artifacts such as paintings and other types of artworks … [that] have the quality of having affordances beyond the instrumental" (Zhu et al., 2017). We have seen the use of daily objects, e.g., lights (Sun et al., 2020a), artwork, e.g., digital paintings (Sun et al., 2017), and mixed reality sandbox, e.g., Inner Garden (Roo et al., 2017), to name a few, to promote a presence-with state of mindfulness. Below, we detail a research-through-design project to showcase how the "total world view" in deep ecology thinking (Naess, 1990) and the "harmonious unity" in Chinese aesthetics can be applied to digital designs for mindfulness.

Case Study on Present-focused Design. "YU" (Zhu et al., 2015, 2017) is an installation of a digital pond with koi (YU means fish in Chinese). In YU, "the movements of fish are animated through isomorphic relations to the physiology of the observer" (Zhu et al., 2017). The pulse of a user can be measured by a sensor attached to the finger or the ear, and the real-time data drives the visual effect of the simulated fish and its Asian-style aquarium. YU intends to help achieve a state of vanishing of the self or no-self through "naturalization of humans" (Zhu et al., 2015) without setting tasks for users to complete. To users, their current physiological signals become the movement of the fish, the shades of its scales, the waves and bubbles in the water, and the water and ambient sound. The mapping can be symmetric or reversed (e.g., stress signals result in calming fish states) and there may exist other mappings that follow or break out of the isomorphic relations to establish a presence-in-nature or presence-with-nature at the current moment.

Time is an important construct in ecology. HCI researchers and designers can exercise ecological thinking in developing and using computing systems with temporal characteristics, whether it is a fast technology under ecological limit, a slow technology for long-term use, a technology inspired by future visions, a technology informed by past history, or a technology focusing on the present moment.

ENTITY

Another key dimension of an ecological system is the entities – artifacts and living things (i.e., human or nonhuman beings) – in it. The field of HCI, as suggested by its name, preliminarily takes a human-centered approach to research and design. In other words, humans are the core of most HCI dialogues. Therefore, in this chapter, we direct our attention to discussions involving other nonhuman creatures and artifacts.

ARTIFACTS

> "The Second Law of Ecology: Everything Must Go Somewhere" – by Barry Commoner in the best-selling book "The Closing Circle: Nature, Man, and Technology"
>
> **(Commoner, 2020)**

Bruno Latour used an analogy to cosmology to illustrate the significance of the world of things around us, describing artifacts – objects, substances, materials, whatever they are called – as "the missing masses of our society … found among the nonhuman mechanisms" (Latour, 1992). Artifacts are a non-dismissible but often neglected part of the ecosystem we live in. Ecological psychologist James J. Gibson introduced the concept of affordance to explain how "the possibilities of the environment and the way of life of the animal go together" (Gibson, 1977). According to him, "the affordances of the environment are what it offers the animal, what it provides or furnishes, either for good or ill" (Gibson, 1977).

Susanne Bødker and Clemens Nylandsted Klokmose proposed an HCI framework of artifact ecology to structure "the understanding of an artifact's action-possibilities" with respect to other things around it (Bødker & Klokmose, 2011). "An artifact ecology often consists of multiple artifacts built for similar purposes, but with slight variations and no clear delineation of when to use which artifact… All artifacts used by human beings are part of artifact ecologies, whether simple (e.g. pen and paper) or complex (e.g. tools for building a house). Hence, human activity is not just mediated through a single artifact; it is multi-mediated" (Bødker & Klokmose, 2011). Edward Reed offered a similar view that humans' "extensive use of tools, transport and modification of materials, and deployment of complex sequences of resource appropriation … put an emphasis on the discovery of relationships among affordances, not just on isolated affordances themselves" (Reed, 1996). The first case study below presents an attempt to disguise physical artifacts with their virtual proxies of similar affordances to maintain a consistence appearance and experience with the surrounding and the ongoing activities.

Case Studies on Artifact Affordance (I). Technologists are expanding our world through extended and hybrid realities (XR). Since the physical space and the virtual space intrinsically overlaps, users need to maintain an awareness of the elements in the actual world as they physically move around in the virtual environment. Whether it is bumping into a chair in virtual reality (VR) or seeing a modern building when trying to interact with AR dinosaurs in city park, users are likely to experience a break in presence (BIP) (He et al., 2018). ARchitect is a proof-of-concept AR-VR prototype that exploits what physical objects have to offer for human activities in the virtual world (Lin et al., 2020). Using ARchitect's AR interfaces, users can scan the environment and identify the objects in it and their common action possibilities (i.e., affordances). The system then prompts a list of possible virtual artifact proxies of the same affordance of a selected object and the users can choose one to cover this physical object in the virtual world (Figure 12.6 (B) and (C)). As shown in Figure 12.6 (A), after going through this process for all key objects, the physical room (b) gets transformed into a virtual farm scene (d) when users put on a VR headset. People can interact with the virtual proxy elements in the same manner as they would act upon the corresponding physical artifacts, which enables them to have a visually consistent experience without worrying about their safety walking around the room.

This first case study exemplifies an HCI design that leverages technologies to connect the physical and virtual experiences with daily objects. In the second case study, we shift our focus to the affordance of computing systems. In Bødker and Klokmose's ecological thinking, computing devices are "more than just objects: They are artifacts, i.e. they are designed or shaped by human beings with a

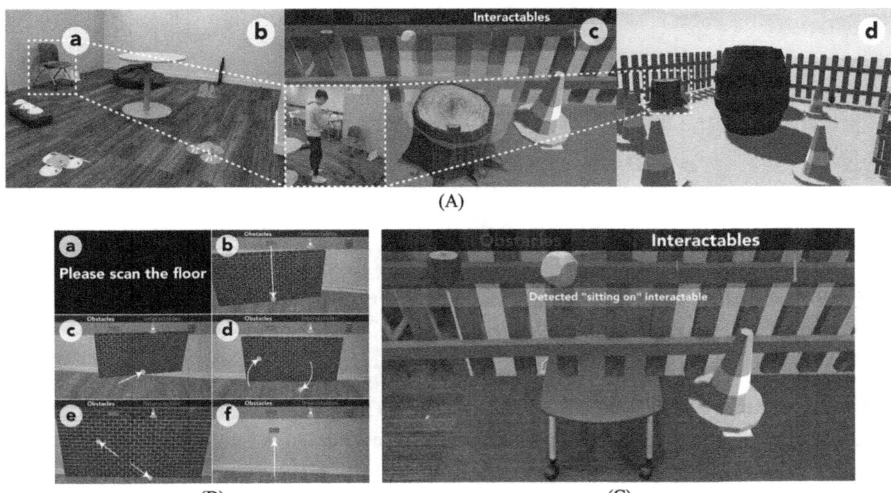

FIGURE 12.6 (A) Workflow of ARchitect: (a) ARchitect maps physical objects to virtual proxies offering matching affordances (e.g., both "brick wall" and "wooden fence" can afford "blocking access"). In this process, objects detected in a physical scene (b) get translated using ARchitect's user interface (c) to an interactive virtual experience (d). (B) Operations for configuring a virtual proxy over the physical scene using ARchitect: (a) scan the floor, (b) place, (c) translate, (d) rotate, (e) resize, or (f) remove the virtual proxy. (C) The Affordance Recommender detects predefined interactable classes (e.g., the "chair" class corresponds to "sitting on" interactable) (Lin et al., 2020).

particular purpose or use in mind" (Bødker & Klokmose, 2011). They pictured "ecology as connected to purposeful, goal-oriented action of some kind, and not as truly endless action possibilities of the environment" (Bødker & Klokmose, 2011). Following this view, we demonstrate that conventional interactive gestures a mobile or wearable device affords can be reframed as meaningful and playful user activities simulated through the device.

Case Studies on Artifact Affordance (II). Metaphoraction is a creativity support tool that enables exploration of designs that reify abstract meanings through common device interactions (Sun et al., 2022). For example, swiping a credit card to make a transaction is a widely adopted interaction. This swiping gesture signifies cutting of a rope to set someone free, communicating the meaning behind the transaction (for a donation). We formulate the design flow for gesture-based interactions to show metaphorical meanings by dissecting the design space into four interconnected components: gesture, action, object, and meaning (Figure 12.7). To this end, we carried out a literature survey to compile a list of widely accepted, lightweight interactive gestures (e.g., tap, swipe, shake, tilt) supported by commercially available or off-the-shelf wearable/mobile devices. We then, conducted a crowdsourced study to translate every interactive gesture in the list into a diverse set of daily actions performed on associated objects by only showing the hand/head/body movements without the presence of the computing device. For instance,

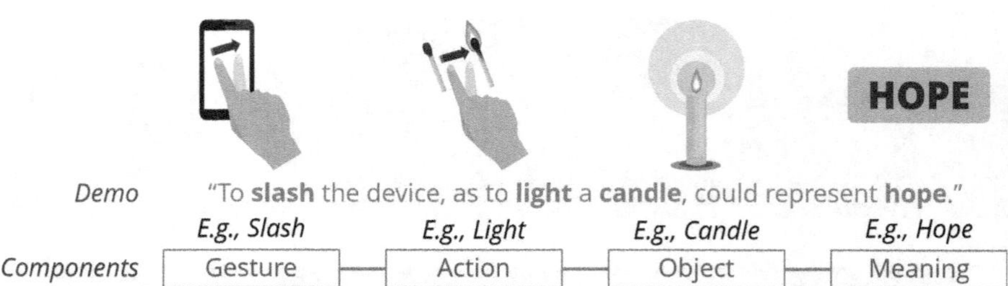

FIGURE 12.7 An overview of Metaphoraction construction with four components: gesture, action, object, and meaning (Sun et al., 2022).

a splash gesture (on the touch screen) assembles striking a match to light a candle. Next, we obtained a large (linguistic) metaphor dataset mined from the Internet to infer possible messages that the objects named in the previous step are likely to convey. For example, candles can represent light, hope, positivity, and goals in a metaphorical sense. We connected the chains of gesture-action-object-meaning form a large space of design ideas. Metaphoraction offers an interactive Sankey diagram visualizing this space and allows users to search for possible candidates with keywords related to one or more of the four components. We presented the system to a professional design team in a workshop and designers explored its use to ideate playful and meaningful gesture-based interactions for their interactive mobile advertisement.

The concept of affordance in HCI concerns the interaction possibilities of a design. Sustainable HCI researchers also tap into technology reuse and repurposing beyond initial acquirement. The second law of ecology "Everything Must Go Somewhere" denotes the fundamental principle in thermodynamics that there is no final waste in nature. All materials in an ecological system get recycled and their identities are constantly transforming in this process. Similarly, values built into a computing system need maintenance and even re-examination for possible extension or transformation. To humans, the world of objects around them – the artifacts – "get designed, purchased, and adopted, but they also get fixed, discarded, and (sometimes) reused" (Jackson & Kang, 2014). This routine practice generates opportunities for "posthumanist alternatives to questions around HCI creativity and design" (Jackson & Kang, 2014). Artists such as those who participated in the "Recology Artist in Residence (AIR) Programs" experiment with the use of unwanted or discarded substances to create diverse forms of art, with the purpose of inspiring the community about new means to achieve resource conservation and sustainability. Technological breakdown and obsolescence can be treated as design materials in a similar manner, serving new needs through creative reuse and repurposing. As Clement of Alexandria put it, "We are not to throw away those things which can benefit our neighbor. Goods are called good because they can be used for good: they are instruments for good, in the hands of those who use them properly." In the following case study, we introduce an exploration in this direction.

Case Study on Artifact Reuse and Repurposing. Project "From Breakage to Icebreaker" set out to explore "why and how accidental breakage of technologies can promote humans to interact and ultimately lead to positive behavioral, emotional, and relational change" (Ma et al., 2016). The project was inspired by the phenomenon that temporary unavailability of technology sometimes creates opportunities for strengthening the interpersonal relationship. For instance, people start to chat when the Wi-Fi in the room is disconnected or when waiting in line for the coffee machine. People even deliberately unplug themselves from the heavy use of technologies and digital media to foster socialization, such as the "phone stacking game" at a group dining event. Following James Pierce's idea of undesigning technology (Pierce, 2012), we sought to "transfer insights from breakage of technology to thoughtful design of technology, ... embedding icebreaking mechanisms into existing products and services to create opportunities for users to interact and reflect while enjoying the original functionalities" (Ma et al., 2016). Through a series of research activities, including anecdote interviews, material-oriented and goal-oriented design workshops, and a case study, we identified design opportunities in restoring, reinforcing, or promoting interpersonal communication, and possible processes to leverage the resourceful surroundings for alternative use. For example, as shown in Figure 12.8, Remotouch (S_2D1) is an air-conditioner remote control design for rooms shared by family members or friends. Remotouch only functions when two people hold it together – a malfunction in a conventional sense. The hand touch and exchange of body temperature serve to remind people to care for each other's feeling, comfort and wellbeing, providing common grounds for negotiation and decision every time they want to adjust the temperature setting.

In a broader sense, the ARchitect and Metaphoraction projects mentioned earlier can be deemed as a reuse and repurpose of artifacts and technologies based on their interactive affordance.

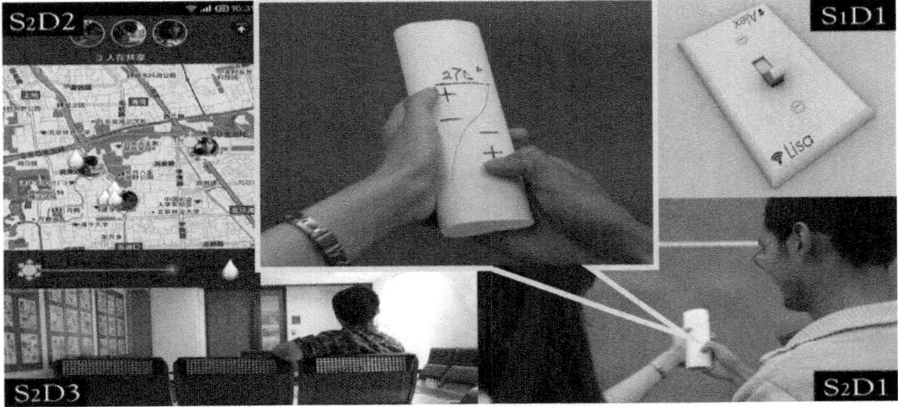

FIGURE 12.8 Four Breakage-to-Icebreaker prototypes generated in the design workshops (Ma et al., 2016).

Nonhuman Living Beings

> The love for all living creatures is the most noble attribute of man.
>
> **– By Charles Darwin**

Humans' view on other nonhuman lives is different from how they position non-living things. When technologies are introduced into our societies, other living beings that co-exist in or are in close proximity to human environments are inevitably exposed to these computing systems or their effects (e.g., occupation of space, reconfiguration of environment, emission of light, sound, heat, and pollutions). Figure 12.9 shows how a cat is drawn to screens of different sizes by the light and moving pictures. In a broader sense, the impact that human activities impose on the ecological environment of the Earth will hit every life cohabitating with us on this planet.

Sustainable HCI researchers are tasked to understand the possible ecological crisis happening to other beings and propose measures to mitigate negative consequences. On the one hand, they inspect the roles technologies have played in the ecological problems; on the other hand, they try to leverage computing systems for the investigation and solution-building. For instance, Patchkeeper is a wearable sensor that can be strapped around a shelter dog's chest to infer its personality from activity data for better matching of potential owners (Meegahapola et al., 2023). The Tingtibi Birdhouse placed in households of remote, sparse-populated communities offers an ambient mode that plays sporadic bird calls at various time intervals and an interactive mode to foster discovery and conservation of endangered species in local wilderness soundscapes (Dema et al., 2020).

Both examples demonstrated the efforts of attending to the wellbeing of nonhuman creatures, domestic or wild. Nevertheless, the former looked at a scenario where living beings other than humans – dogs in this case – are the solo or key users of computing systems. Clara Mancini wrote an article on animal-computer interaction, calling for the development interactive technologies that "improve animals' life expectancy and quality by facilitating the fulfillment of their physiological

FIGURE 12.9 My cat stared at a phone, a tablet, a laptop, and a TV screen. (Copyright @ Xiaojuan Ma.)

and psychological needs, ..., support animals in the legal functions..., and foster the relationship between humans and animals by enabling communication and promoting understanding between them" (Mancini, 2011). The article highlighted the ethical considerations in this line of work, which emphasized that both human and nonhuman participants should be respected, treated, and cared equally. In an interesting HCI project, researchers tried to answer the question "Would parrots freely engage in making video calls to other parrots?" through a two-phased, longitudinal study (Kleinberger et al., 2023). In the first phase, the participating parrots were trained to have video calls with one another with a treat. In Phase 2, the birds had the agency to request video calls by pressing the photo of another bird on their tablet and no treats were provided. Such an experience is positioned as "a form of enrichment that should fit in the animal's existing social and living context" (Kleinberger et al., 2023).

Some scholars advocate for an even more radical approach to disentangle the different members of cohabitation (Smith et al., 2017) and place nonhuman lives in the foreground of the picture, i.e., designing "to unseat humans from the center of the universe and support a more equitable gaze" (Light et al., 2017). This is an actualization of ecological posthumanism and multi-species thinking. In Timothy Morton's ecological view, what we call nature is "a vast, sprawling mesh of interconnection without a definite center or edge" and is "a radical intimacy, coexistence with other beings, sentient and otherwise" (Morton, 2010). To him, ecological thought should seek to enchant the world, "exploring the profound and wonderful openness and intimacy of the mesh" (Morton, 2010). As a result, the ecological thought is "full of shadows and twilights" and the ecological world "isn't a positive, sunny ... world" (Morton, 2010). Robin Kimmerer shared the same insight and called "our response to the generosity of the more-than-human world" the act of "paying attention" (Kimmerer, 2014). As Kimmerer noted, "Paying attention to the more-than-human world doesn't lead only to amazement; it leads also to acknowledgment of pain. Open and attentive, we see and feel equally the beauty and the wounds, the old growth and the clear-cut, the mountain and the mine. Paying attention to suffering sharpens our ability to respond. To be responsible" (Kimmerer, 2014).

The practice of paying attention gets translated into an HCI research and design method called "arts of noticing" (Liu et al., 2019b). This term is inherited from anthropologist Anna Tsing who promoted revisiting research assumptions to uncover alternative paths heading toward preferable futures (Tsing, 2015). HCI scholars have extended the notion of noticing and practice it as a means to cultivate "alternative perspectives in technological intervention" and "engage with the challenges presented by intersecting environmental and social crises" (Liu et al., 2019b). For example, sustainable agriculture explores alternative farming techniques that are less deleterious to the environment and a group of HCI researchers conducted ethnographic studies on selected sites to learn about such new human-land relationships (Liu et al., 2019a). They looked into "nontraditional users (e.g., nonhumans) and emerging forms of uses (e.g., interactions between human and other species) to help open a design space for technological interventions" (Liu et al., 2019a). At the same front, researchers from the same group exercised the arts of noticing in an autoethnographic bird watching activity (Biggs et al., 2021).

They "honed a practice of attunement through deep listening and field recording," found "ways to build intimacy with birds and 'lose our edges'," and gained immersive "ecological" experiences (Biggs et al., 2021). This echoes Morton's ecological thoughts and demonstrates the ability of noticing to inspire HCI scholars to challenge the human/nonhuman boundary and decentralize humans in their pursuit of a more sustainable and desirable future.

CONCLUSION

This chapter synthesizes various schools of ecological thinking and their use in the field of HCI at different levels. We presented different working definitions of ecology and assorted ecological theories; some are complementary while others are competing. We also discuss how the ecological aspects in HCI relate to another notion in HCI – sustainability. We then dive into three key dimensions of ecological thinking – "space," "time," and "entity" – and use case studies to illustrate how the corresponding branch of ecological thinking can inform HCI research and design.

REFERENCES

Barba, E., & MacIntyre, B. (2011, November). A scale model of mixed reality. In *Proceedings of the 8th ACM Conference on Creativity and Cognition* (pp. 117–126). ACM.

Biggs, H. R., Bardzell, J., & Bardzell, S. (2021, May). Watching myself watching birds: Abjection, ecological thinking, and posthuman design. In *Proceedings of the 2021 CHI Conference on Human Factors in Computing Systems* (pp. 1–16).

Blevis, E., Bødker, S., Flach, J., Forlizzi, J., Jung, H., Kaptelinin, V., Nardi, B., & Rizzo, A. (2015, April). Ecological perspectives in HCI: Promise, problems, and potential. In *Proceedings of the 33rd Annual ACM Conference Extended Abstracts on Human Factors in Computing Systems* (pp. 2401–2404).

Bødker, S., & Klokmose, C. N. (2011). The human–artifact model: An activity theoretical approach to artifact ecologies. *Human–Computer Interaction, 26*(4), 315–371.

Bourdieu, P. (1990). *The logic of practice.* Stanford University Press.

Buçinca, Z., Malaya, M. B., & Gajos, K. Z. (2021). To trust or to think: Cognitive forcing functions can reduce overreliance on AI in AI-assisted decision-making. *Proceedings of the ACM on Human-Computer Interaction, 5*(CSCW1), 1–21.

Cervato, C., & Frodeman, R. (2012). The significance of geologic time: Cultural, educational, and economic frameworks. *Geological Society of America Special Papers, 486*, 19–27.

Code, L. (2006). *Ecological thinking: The politics of epistemic location.* Oxford University Press.

Commoner, B. (2020). *The closing circle: Nature, man, and technology.* Courier Dover Publications.

Conley, V. A. (2006). *Ecopolitics: The environment in poststructuralist thought.* Routledge.

Daniel, K. (2017). Thinking, fast and slow. Farrar, Straus and Giroux. New York.

Deleuze, G. (1988). *Spinoza: Practical philosophy.* City Lights Books.

Deleuze, G., & Guattari, F. (1994). *What is philosophy?* Columbia University Press.

Dema, T., Brereton, M., Esteban, M., Soro, A., Sherub, S., & Roe, P. (2020, April). Designing in the network of relations for species conservation: The playful Tingtibi community birdhouse. In Proceedings of the 2020 CHI Conference on Human Factors in Computing Systems (pp. 1–14).

Fletcher, R., Fortin, M. J., Fletcher, R., & Fortin, M. J. (2018). Introduction to spatial ecology and its relevance for conservation. *Spatial Ecology and Conservation Modeling: Applications with R*, 1–13.

Forlizzi, J. (2008). The product ecology: Understanding social product use and supporting design culture. *International Journal of Design*, *2*(1).

Fuller, M. (2005). *Media ecologies: Materialist energies in art and technoculture*. MIT Press.

Gibson, J. J. (1977). The theory of affordances. *Hilldale, USA*, *1*(2), 67–82.

Hallnäs, L., & Redström, J. (2001). Slow technology–designing for reflection. *Personal and ubiquitous computing*, *5*, 201–212.

Hayden, P. (1997). Gilles Deleuze and naturalism: A convergence with ecological theory and politics. *Environmental Ethics*, *19*(2), 185–204.

Hazas, M., & Nathan, L. (Eds.). (2017). *Digital technology and sustainability: Engaging the paradox*. Routledge.

Heringman, N. (2015). Deep time at the dawn of the Anthropocene. *Representations*, *129*(1), 56–85.

He, Z., Zhu, F., Perlin, K., & Ma, X. (2018). Manifest the invisible: Design for situational awareness of physical environments in virtual reality. arXiv preprint arXiv: 1809.05837.

Irvine, R. D. G. (2014). Deep time: An anthropological problem. *Social Anthropology*, *22*(2), 157–172.

Jackson, S. J., & Kang, L. (2014, April). Breakdown, obsolescence and reuse: HCI and the art of repair. In *Proceedings of the SIGCHI conference on human factors in computing systems* (pp. 449–458).

Jacob, R. J., Girouard, A., Hirshfield, L. M., Horn, M. S., Shaer, O., Solovey, E. T., & Zigelbaum, J. (2008, April). Reality-based interaction: A framework for post-WIMP interfaces. In *Proceedings of the SIGCHI conference on Human factors in computing systems* (pp. 201–210).

Kaptelinin, V. (2014). *Affordances and design*. Interaction Design Foundation.

Kimmerer, R. W. (2014). Returning the gift. *Minding Nature*, *7*(2), 18–24.

Kingsland, S. E. (1995). *Modeling nature*. University of Chicago Press.

Kjærup, M., Skov, M. B., Nielsen, P. A., Kjeldskov, J., Gerken, J., & Reiterer, H. (2021). *Longitudinal studies in HCI research: A review of CHI publications from 1982–2019* (pp. 11–39). Springer International Publishing.

Kleinberger, R., Cunha, J., Vemuri, M. M., & Hirskyj-Douglas, I. (2023, April). Birds of a feather video-flock together: Design and Evaluation of an agency-based parrot-to-parrot video-calling system for interspecies ethical enrichment. In *Proceedings of the 2023 CHI Conference on Human Factors in Computing Systems* (pp. 1–16).

Knowles, B., Bates, O., & Håkansson, M. (2018). This changes sustainable HCI. In *Proceedings of the CHI 2018* (pp. 1–12). Montréal, QC, Canada: ACM.

Latour, B. (1992). Where are the missing masses? The sociology of a few mundane artifacts, in Bijker, W. E. and Law, J. (Eds.), *Shaping Technology/Building Society: Studies in Sociotechnical Change* (pp. 225–58), MIT Press.

Lear, L. (1998). *Rachel Carson: Witness for nature*. Macmillan.

Light, A., Shklovski, I., & Powell, A. (2017, May). Design for existential crisis. In *Proceedings of the 2017 CHI Conference Extended Abstracts on Human Factors in Computing Systems* (pp. 722–734).

Lin, C. E., Cheng, T. Y., & Ma, X. (2020, April). Architect: Building interactive virtual experiences from physical affordances by bringing human-in-the-loop. In *Proceedings of the 2020 CHI Conference on Human Factors in Computing Systems* (pp. 1–13).

Liu, S. Y., Bardzell, S., & Bardzell, J. (2019a). Symbiotic encounters: HCI and sustainable agriculture. In *Proceedings of the 2019 CHI Conference on Human Factors in Computing Systems* (pp. 1–13).

Liu, S. Y., Liu, J., Dew, K., Zdziarska, P., Livio, M., & Bardzell, S. (2019b). Exploring noticing as method in design research. In *Companion Publication of the 2019 on Designing Interactive Systems Conference 2019 Companion* (pp. 377–380).

Ma, X., Fang, K., & Zhu, F. (2016, June). From breakage to icebreaker: Inspiration for designing technological support for human-human interaction. In *Proceedings of the 2016 ACM Conference on Designing Interactive Systems* (pp. 403–414).

Mancini, C. (2011). Animal-computer interaction: A manifesto. *Interactions, 18*(4), 69–73.

Mazé, R., & Redström, J. (2005). Form and the computational object. *Digital Creativity, 16*(1), 7–18.

Meegahapola, L., Constantinides, M., Radivojevic, Z., Li, H., Quercia, D., & Eggleston, M. S. (2023, April). Quantified canine: Inferring dog personality from wearables. In *Proceedings of the 2023 CHI Conference on Human Factors in Computing Systems* (pp. 1–19).

Montello, D. R. (1993, September). Scale and multiple psychologies of space. In *European Conference on Spatial Information Theory* (pp. 312–321). Springer Berlin Heidelberg.

Montero, C. S., Alexander, J., Marshall, M. T., & Subramanian, S. (2010, September). Would you do that? Understanding social acceptance of gestural interfaces. In *Proceedings of the 12th International Conference on Human Computer Interaction with Mobile Devices and Services* (pp. 275–278).

Morton, T. (2010). *The ecological thought.* Harvard University Press.

Naess, A. (1990). *Ecology, community and lifestyle: Outline of an ecosophy.* Cambridge University Press.

Naess, A. (2017). The shallow and the deep, long-range ecology movement. A summary. In *The ethics of the environment* (pp. 115–120). Routledge.

Nardi, B. A., & O'Day, V. (2000). *Information ecologies: Using technology with heart.* MIT Press.

Norman, D. (2004). Affordances and design. Unpublished article, available online at: https://www.researchgate.net/publication/265618710_Affordances_and_Design

Norman, D. A. (1988). *The psychology of everyday things.* Basic Books.

Norman, D. A. (1990). *The design of everyday things.* Doubleday.

Odom, W. (2015, April). Understanding long-term interactions with a slow technology: An investigation of experiences with FutureMe. In *Proceedings of the 33rd Annual ACM Conference on Human Factors in Computing Systems* (pp. 575–584).

Odom, W., Linehan, C., Pschetz, L., Tsaknaki, V., Vallgårda, A., Wiberg, M., & Yoo, D. (2018a). Time, temporality, and slowness: Future directions for design research. In *Proceedings of the 2018 ACM Conference Companion Publication on Designing Interactive Systems* (pp. 383–386). Hong Kong, China: ACM.

Odom, W. T., Sellen, A. J., Banks, R., Kirk, D. S., Regan, T., Selby, M., Forlizzi, J. L., & Zimmerman, J. (2014, April). Designing for slowness, anticipation and re-visitation: A long term field study of the photobox. In *Proceedings of the SIGCHI Conference on Human Factors in Computing Systems* (pp. 1961–1970).

Odom, W., Wakkary, R., Bertran, I., Harkness, M., Hertz, G., Hol, J., Lin, H., Naus, B., Tan, P., & Verburg, P. (2018b). Attending to slowness and temporality with Olly and slow game: A design inquiry into supporting longer-term relations with everyday computational objects. In *Proceedings of the 2018 CHI Conference on Human Factors in Computing Systems* (pp. 1–13).

Odom, W., Wakkary, R., Hol, J., Naus, B., Verburg, P., Amram, T., & Chen, A. Y. S. (2019, May). Investigating slowness as a frame to design longer-term experiences with personal data: A field study of Olly. In *Proceedings of the 2019 CHI Conference on Human Factors in Computing Systems* (pp. 1–16).

Park, J. S., Barber, R., Kirlik, A., & Karahalios, K. (2019). A slow algorithm improves users' assessments of the algorithm's accuracy. *Proceedings of the ACM on Human-Computer Interaction, 3*(CSCW), 1–15.

Peng, Z., Mo, K., Zhu, X., Chen, J., Chen, Z., Xu, Q., & Ma, X. (2020, April). Understanding user perceptions of robot's delay, voice quality-speed trade-off and GUI during conversation. In *Extended Abstracts of the 2020 CHI Conference on Human Factors in Computing Systems* (pp. 1–8).

Pierce, J. (2012, May). Undesigning technology: Considering the negation of design by design. In *Proceedings of the SIGCHI Conference on Human Factors in Computing Systems* (pp. 957–966).

Rahm-Skågeby, J., & Rahm, L. (2022). HCI and deep time: Toward deep time design thinking. *Human–computer interaction*, *37*(1), 15–28.

Reed, E. S. (1996). *Encountering the world: Toward an ecological psychology.* Oxford University Press.

Rico, J., & Brewster, S. (2010, April). Usable gestures for mobile interfaces: Evaluating social acceptability. In *Proceedings of the SIGCHI Conference on Human Factors in Computing Systems* (pp. 887–896).

Roo, J. S., Gervais, R., Frey, J., & Hachet, M. (2017, May). Inner garden: Connecting inner states to a mixed reality sandbox for mindfulness. In *Proceedings of the 2017 CHI Conference on Human Factors in Computing Systems* (pp. 1459–1470).

Shekhar, S., Feiner, S. K., & Aref, W. G. (2015). Spatial computing. *Communications of the ACM*, *59*(1), 72–81.

Shekhar, S., & Vold, P. (2020). *Spatial computing.* MIT Press.

Smith, N., Bardzell, S., & Bardzell, J. (2017, May). Designing for cohabitation: Naturecultures, hybrids, and decentering the human in design. In *Proceedings of the 2017 CHI Conference on Human Factors in Computing Systems* (pp. 1714–1725).

Sun, Z., Cao, N., & Ma, X. (2017, May). Attention, comprehension, execution: Effects of different designs of biofeedback display. In *Proceedings of the 2017 CHI Conference Extended Abstracts on Human Factors in Computing Systems* (pp. 2132–2139).

Sun, Z., Han, F., & Ma, X. (2018, April). Exploring the Effects of Scale in Augmented Reality-Empowered Visual Analytics. In *Extended Abstracts of the 2018 CHI Conference on Human Factors in Computing Systems* (pp. 1–6).

Sun, Z., Reani, M., Li, Q., & Ma, X. (2020a). Fostering engagement in technology-mediated stress management: A comparative study of biofeedback designs. *International Journal of Human-Computer Studies*, *140*, 102430.

Sun, Z., Wang, S., Liu, C., & Ma, X. (2022). Metaphoraction: Support gesture-based interaction design with metaphorical meanings. *ACM Transactions on Computer-Human Interaction* (TOCHI), *29*(5), 1–33.

Sun, Z., Wang, S., Yang, W., Yürüten, O., Shi, C., & Ma, X. (2020b). "A Postcard from Your Food Journey in the Past": Promoting Self-Reflection on Social Food Posting. In *Proceedings of the 2020 ACM Designing Interactive Systems Conference* (pp. 1819–1832).

Tilman, D., & Kareiva, P. (Eds.). (1997). *Spatial ecology: The role of space in population dynamics and interspecific interactions.* Princeton University Press.

Tsing, A. L. (2015). *The mushroom at the end of the world: On the possibility of life in capitalist ruins.* Princeton University Press.

Tversky, B. (2003). Structures of mental spaces: How people think about space. *Environment and behavior*, *35*(1), 66–80.

Vaughan, M., Courage, C., Rosenbaum, S., Jain, J., Hammontree, M., Beale, R., & Welsh, D. (2008). Longitudinal usability data collection: Art versus science? In *CHI'08 extended abstracts on Human factors in computing systems* (pp. 2261–2264).

Williams, R. (1976). *Keywords: A vocabulary of culture and society.* Oxford University Press.

Wu, Z., Guo, J., Zhang, S., Zhao, C., & Ma, X. (2019, July). An AR benchmark system for indoor planar object tracking. In *2019 IEEE International Conference on Multimedia and Expo* (ICME) (pp. 302–307). IEEE.

Wu, C., He, K., Chen, J., Zhao, Z., & Du, R. (2020). Liveness is not enough: Enhancing fingerprint authentication with behavioral biometrics to defeat puppet attacks. In *29th USENIX Security Symposium (USENIX Security 20)* (pp. 2219–2236).

Zhao, M., Chen, Z., Lu, K., Li, C., Qu, H., & Ma, X. (2016, October). Blossom: Design of a tangible interface for improving intergenerational communication for the elderly. In *Proceedings of the International Symposium on Interactive Technology and Ageing Populations* (pp. 87–98).

Zhu, B., Hedman, A., & Li, H. (2017). Designing digital mindfulness: Presence-in and presence-with versus presence-through. *In Proceedings of the 2017 CHI Conference on Human Factors in Computing Systems (CHI'17)* (pp. 2685–2695). ACM.

Zhu, Q., Wang, Z., Zeng, W., Tong, W., Lin, W., & Ma, X. (2024). Make Interaction Situated: Designing User Acceptable Interaction for Situated Visualization in Public Environments. In Proceedings of the CHI Conference on Human Factors in Computing Systems (pp. 1–21).

Zhu, B., Zhang, Y., Ma, X., & Li, H. (2015). Bring Chinese Aesthetics into Designing the Experience of Personal Informatics for Wellbeing. In 9th Interactional Conference on Design and Semantics of Form and Movement (DeSForM 2015).

13 Social HCI

Zhicong Lu

THE RISE OF SHORT VIDEOS AND LIVE STREAMING

Online communities have evolved as integral components of our digital lives (Preece, 2000). These virtual spaces, formed around shared interests, experiences, or objectives, have transformed how people interact, learn, and influence each other online. In the last few decades, online communities have evolved from text-based forums to more interactive environments, including image-based social media such as Instagram, virtual worlds in video games, and more recently, platforms dedicated to creating and sharing short videos and live streams such as TikTok and Twitch. These new avenues of sharing user-generated content online, with their unique affordances, have notably altered the dynamics of online community building and user engagement.

In the past decade, the rise of short videos and live streaming has dramatically shifted the landscape of social media and online communities. For example, one of the most popular short-video platforms, TikTok, has over 1 billion monthly active users worldwide (Lu et al., 2020; Lu & Lu, 2019). The platform's rapid growth and widespread adoption have made it a significant player in the social media landscape. This global trend has not only revolutionized the way online content is produced and consumed but also brought about novel challenges and opportunities for HCI research (Hamilton et al., 2014; Lu, Xia, et al., 2018b; Tang et al., 2016).

The ubiquity of mobile devices equipped with high-definition cameras and high-speed Internet has fostered a user-centric multimedia universe where anyone can become a content creator (Lu, Xia, et al., 2018b; Tang et al., 2016). Short-video platforms like TikTok and live streaming services such as Twitch and Douyu have democratized content creation and diversified the media landscape. Moreover, the real-time nature of live streaming creates a unique and interactive social experience, bridging the gap between content creators and consumers. Content creators produce short videos or live streams of a wide range of genres, such as video gaming (Hamilton et al., 2014), live events (Haimson & Tang, 2017), civil content (Dougherty, 2011), live performances (Lin & Lu, 2017), or the selling of goods (Cai et al., 2018; Wu et al., 2023). Short videos and live streams enable different online communities to form and sustain, facilitating real-time, synchronous inter-personal interactions between different users.

The HCI challenges presented by these evolving short-video and live streaming platforms predominantly revolve around social interactions, content creation, dissemination, and user engagement. These platforms must offer intuitive, user-friendly user interfaces that cater to a variety of user needs including seamless navigation, personalization, real-time interactions, and accessibility. As users increasingly become content creators, including *professionals* who live on creating

DOI: 10.1201/9781032693606-17

online content, it is crucial to understand how to design, structure, and manage these platforms to support their creative processes. For instance, HCI can contribute to providing mechanisms for users to creatively capture, edit, and publish user-generated content, as well as building inclusive and safe peer-support communities among content creators.

The dissemination of content is another important aspect of short-video and live streaming platforms. These platforms often have congested spaces with limited user attention, making it challenging for creators to have their content discovered by viewers. To address this, the design of recommendation algorithms, integration of social networks, and an understanding of digital echo chambers are all of importance. Additionally, bridging the gap between creators and their fans through real-time interactions on live streams or comments on short videos, also requires considerable design and management, because many undesirable behaviors may emerge.

Understanding and addressing these grand challenges require an interdisciplinary framework that incorporates user-centered design, cognitive psychology, sociology, communication, media studies, information management, and computer science. This makes social HCI an exciting field of discovery.

LIVE STREAMING AND SHORT VIDEO FOR EMPOWERING MARGINALIZED COMMUNITIES

The impact of short-video and live streaming platforms extends beyond entertainment and social interaction, offering significant potential for supporting traditionally under-resourced and marginalized groups, such as traditional cultural practitioners, ethnic minorities, rural users, and aging people.

For example, these short-video and live streaming platforms have become empowering tools in preserving and promoting intangible cultural heritage (ICH), such as folklore, customs, beliefs, and traditions of many local communities, especially in China. A study by Lu et al. (2019) demonstrated how livestreaming can be used to document and share traditional cultural practices, often threatened by globalization and modernization. Through livestreaming, these traditional cultural practices can reach a broader audience, ensuring their preservation and promoting cultural diversity. The study shows that Chinese ICH content creators often demonstrate altruistic motivations to share ICH-related content via live streams and short videos. They engage with viewers using multiple modalities and channels, such as WeChat group chats. Live streaming and short video also complement each other when supporting content creators: livestreaming encourages real-time interaction and sociality, while non-live curated short videos attract attention from a broader audience and assist in archiving ICH knowledge. The study also highlights several challenges associated with ICH live streaming and content creation, such as the fragmented technology eco-system, the physical and cognitive demands of live streaming complex activities, and viewers' misunderstandings and unintentional trolling that discourage ICH content creators.

Similarly, another study by Chen et al. (2023) delves into ethnic minorities' lived experiences in China, who face challenges in sustaining their own culture in regions dominated by ethnic majorities. With the growing popularity of video blogging (vlogging) on short-video platforms in China, many ethnic minority vloggers are using vlogs to present and promote their ethnic culture on Douyin. The study shows that both ethnic cultural experts and non-experts are involved in ethnic vlog making and sharing. Cultural experts take more initiative in preserving and promoting ethnic culture, while non-experts are more motivated by getting more traffic and income. Vloggers' imagined audiences include both intra-ethnic and mainstream viewers, impacting their vlog-making strategies, the utilized platform features, and the created vlog content. For example, some vloggers teach ethnic languages and build a unique identity for intra-ethnic viewers. By supporting social interactions and conversations, Douyin enables ethnic minority vloggers and viewers to collaboratively protect their culture from misinterpretation by mainstream viewers.

E-commerce livestreaming has also proven to be a transformative tool in empowering Chinese rural women. A study by Tang et al. (2022) elucidated how livestreaming empowers rural women in China by providing them with new employment opportunities, fostering entrepreneurship, and enhancing their business visibility. This study shows that e-commerce live streaming democratizes the digital space, enabling marginalized groups to participate actively in the e-commerce boom in China. These rural women gain a sense of self-empowerment through economic, social, intellectual, psychological, and spiritual dimensions. However, they also experience and confront social stigmas rooted in rural societies. For example, undesirable gender norms were still rife in rural China, i.e., female live streamers were considered to be the ones who created sexual imagery online to attract male viewers. Despite the empowerment that e-commerce live streaming may bring, rural female live streamers still face a lot of social pressure when conducting e-commerce live streaming. Interestingly, some female live streamers seek endorsement from local government in order to conduct live streaming without harming their reputations.

Further, these platforms also provide potential romantic relationship opportunities or job opportunities for aging people, an often overlooked demographic. A recent study by C. He et al. (2023) explored the use of streamer-moderated matchmaking among single seniors in China. The study found that compared to algorithm-mediated online dating, live streaming-based matchmaking is characterized by the mediation of matchmakers in a synchronous virtual environment and the natural development of live streaming-based matchmaking communities. The study involved observations and semi-structured interviews with six livestreaming matchmakers and twelve senior match-seekers aged 50–70. The findings revealed that matchmakers play a crucial role in facilitating seniors' match-seeking during and beyond livestreaming, and they also have additional duties to enhance seniors' safety in the process. Livestreaming-based matchmaking communities provide multiple important values for single seniors to acquire companionship and help in seeking late-life love. The study unpacked the perceived benefits and challenges of livestreamer-moderated matchmaking and discussed how to support single seniors' match-seeking in an accessible, safe, and convenient manner. It also explored the use of live-streamer-moderated matchmaking among single seniors in China. The study shows that compared to algorithm-mediated online dating, livestreaming-based matchmaking is characterized by the mediation of matchmakers in a synchronous virtual environment and the natural development of livestreaming-based matchmaking communities. The findings reveal that matchmakers play a crucial role in facilitating seniors' match-seeking during and beyond live streaming, and they also have additional duties to enhance seniors' safety in the process. Livestreaming-based matchmaking communities provide multiple important values for single seniors to seek companionship and late-life love.

Another study by Wang et al. (2024) shows how short-video and live streaming platforms cater to aging job seekers' unique needs. Instead of leveraging existing online platforms that specialize in supporting job-seeking (e.g., LinkedIn), a notable number of aging job seekers in China are using short-video and livestreaming platforms to seek jobs, even though these platforms are not initially designed for supporting job-seeking. These platforms mitigate the challenge of limited digital literacy and provide vivid presentations of job opportunities. Additionally, they safeguard the fundamental welfare and rights of aging job seekers by monitoring the qualifications and public commitments made by recruiters targeting aging workers in their videos.

These studies underscore the potential of short-video and live streaming platforms in bridging digital and socio-economic gaps, offering new opportunities to traditionally under-resourced and marginalized communities. This highlights the essential role that HCI research plays in understanding and enhancing the user experience for these communities. It further emphasizes the need for designing interfaces and functionalities that are inclusive and accessible to a diverse user base.

INCREASING USER ENGAGEMENT VIA SUPPORTING RICHER SOCIAL INTERACTIONS

Prior research in live streaming has shown that interactions and the sociality afforded by live streaming are what make it engaging, and both streamers and viewers desire deeper and more meaningful interactions in live streaming beyond just text-based chat (Hamilton et al., 2014; Lu et al., 2019; Lu, Xia, et al., 2018b).

To enable streamers to better engage their viewers, several research projects have explored novel interfaces to facilitate audience participation and influence in video game live streaming. For example, Lessel et al. (2017) designed several new communication channels for live streams of card game Hearthstone, for example, enabling viewers to draw lines on the game UI to indicate their preferred player act and recommend this act to the streamer. Through a preliminary user study, they found that these additional communication channels were favored because they enabled viewers to exert their influence on streamers. However, this system is designed specifically for the game Hearthstone, and the features may not work well for other games or live streaming genres.

Glickman et al. (2018) explored design challenges of live streamed audience participation games (APGs) which enabled viewers to participate in a streamer's gameplay and influence the gaming experience of the streamer. They identified six design challenges of APGs, including latency, screen sharing, attention management, player agency, audience-streamer relationships, and shifting schedules. Their findings were mostly constrained in the domain of videogaming live streams, and the design implications could not be easily generalizable to other genres of live streams.

Although live streaming has been increasingly used in non-gaming contexts, few HCI systems or tools have been designed to better engage viewers in these contexts, largely due to the lack of empirical studies which reveal non-gaming live streaming users' unique needs and practices.

Some exceptions have explored enriched communication channels in other non-gaming genres of live streams, including live events, public environment, education, and knowledge sharing. For instance, LiveMâché (Hamilton et al., 2018) proposed the concept of collaborative live media curation, which supported sharing context and participation in online learning through the collaborative and synchronous curation of multi-media, including live video, text, images, and sketches. Users can freely compose different modalities of media in a curation when attending a live session, and engage with different modality of information to learn. However, this method may not scale to many viewers, because users can be easily overloaded by the rich information curated in the system.

Reeves et al. (2015) studied a mixed reality game which supported live streaming from streets and identified challenges facing streamers such as demands from remote viewers, situational challenges, and self-presentation desires. Chen et al. (2019) explored the use of text, audio, video, image, and stickers to enrich learner interactions for live streaming in language learning, and found that these multimedia tools increased learners' engagement. Snapstream (Yang et al., 2020) enabled viewers to take snapshots of live streams for visual art in real time, annotate on them, and share the annotated snapshots with other users.

StreamWiki (Lu, Heo, et al., 2018a) is a collaborative note-taking system designed to enable viewers of knowledge-sharing live stream to collectively generate a structured archive of the live stream in real-time (Figure 13.1). It serves as an example of how HCI can facilitate active viewer participation, increase information retention, and provide a comprehensive overview of the live-streamed content. In this respect, future work can expand on such interactive and collaborative note-taking systems for livestreaming, exploring how they can be made more effective, user-friendly, and inclusive.

Meanwhile, StreamSketch is an illustrative interaction tool for creative live streaming (Lu et al., 2021). It enables viewers and streamers to interact during live streams using multiple modalities, including freeform sketches and text (Figure 13.2). It reimagines how audiences engage with and process live-streamed content. The development of such tools opens up new avenues for HCI research, particularly around understanding how audiences interact with these tools, how to enhance their usability, and how they can be utilized across diverse contexts. Both StreamWiki and StreamSketch contribute to transforming passive viewers into active participants, making live streaming a more interactive, participatory, and engaging experience.

Prior work has shown that multi-modal interactions in live streaming are feasible, favorable, engaging, and beneficial to both the streamers and viewers. To continue

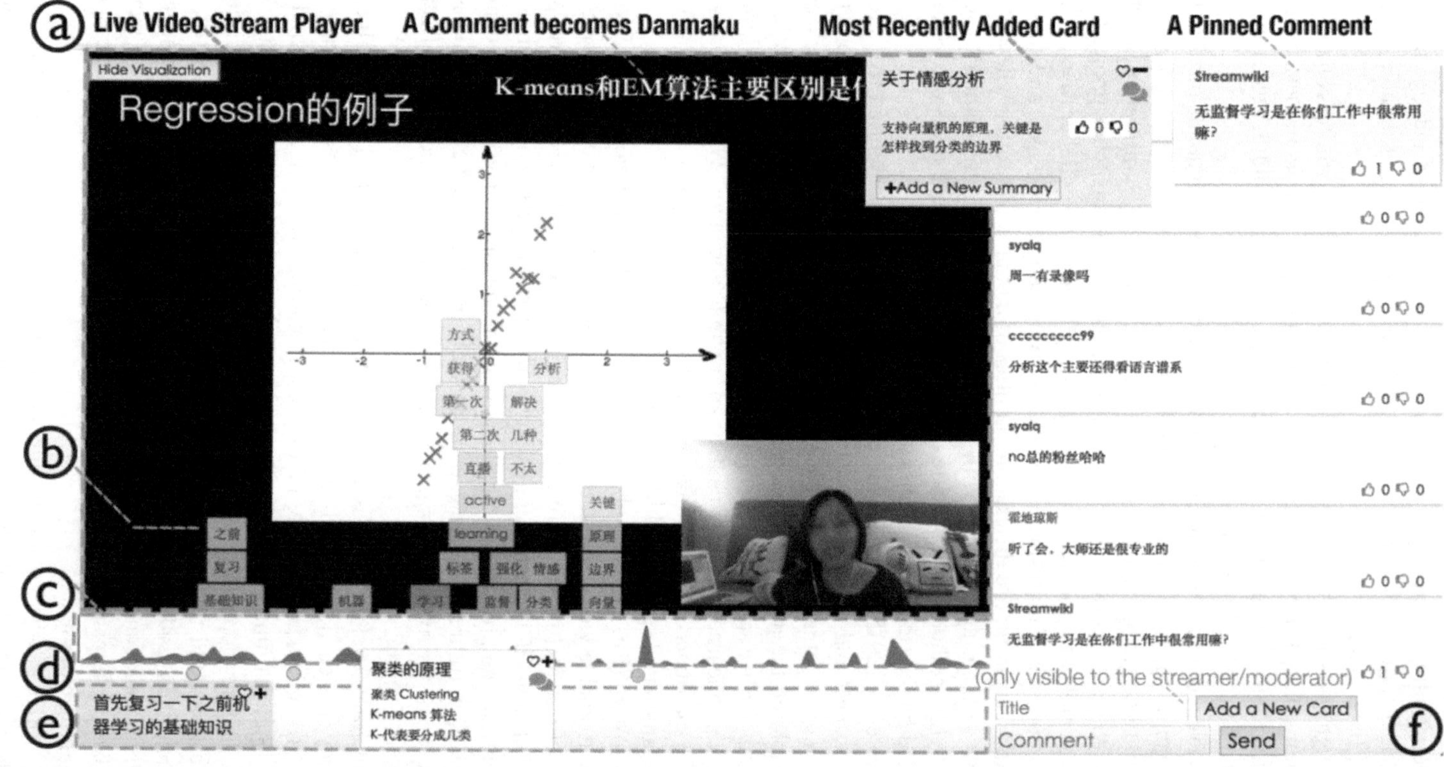

FIGURE 13.1 User interfaces of StreamWiki (Lu, Heo, et al., 2018).

FIGURE 13.2 User interfaces of StreamSketch (Lu et al., 2021).

this line of research, future HCI research can focus on providing richer interaction modalities to streamers, viewers, and other stakeholders, at the same time, reducing information overload and context collapse caused by the rich information from a large number of viewers watching live streams, so that live streaming user interfaces can cater to varied user needs and contexts.

CONTENT MODERATION AND CURATION: DEMYSTIFYING AND MITIGATING THE DARK SIDE OF SOCIAL MEDIA

The rise of negative content in online communities has become a disturbing norm (Gillespie, 2018; Roberts, 2019; Seering et al., 2017). The prevalence of such negative content, which includes hate speech, cyberbullying, disinformation, and disturbing or offensive imagery, can be attributed to several factors. Short-video and live streaming platforms are also increasingly impacted by such behaviors.

The impersonal nature of digital communication, the anonymity that many platforms afford, and the thrill of pushing societal norms, often empower users to create and distribute toxic content (Roberts, 2019). The lack of stringent checks, robust monitoring systems, and swift in-time punitive measures further contribute to the problem. Moreover, short-video and live streaming platforms, due to their widespread reach, real-time interaction, and fast content creation dynamics, can inadvertently facilitate the spread of this negative content and negative user behaviors (Seering et al., 2017; Wohn, 2019).

The impact of such negative content on users ranges from fostering a culture of fear and hostility to psychological harm, which can lead to serious real-world implications, including mental health issues and polarization of society (Roberts, 2019). Furthermore, the reputation of the platforms that host such content can also be severely damaged, leading to user attrition and potential legal consequences (Gillespie, 2018).

In the context of content moderation, the biggest challenge lies in managing the large volume of user-generated content. In real-time ecosystems such as live streaming platforms, this challenge is particularly daunting. While automatic content detection and removal systems can be employed, they often struggle to accurately determine the context, leading to both false positives and negatives. Manual moderation by human moderators, on the other hand, can be emotionally taxing for moderators and is not entirely scalable given the volume of content generated on live streaming platforms (Wohn, 2019).

Because live streaming platforms support more complicated social interaction compared to text-based platforms, new forms of negative behaviors can also easily emerge within live streaming. For example, a recent study examines the phenomenon of *hate raids* on Twitch (Han et al., 2023). The study uses a mixed-methods approach, combining quantitative measurement of attacks across the platform with interviews of streamers and third-party bot developers. The study found that hate raids on Twitch were highly targeted, hate-driven attacks that disproportionately targeted streamers who self-identify as LGBTQ+ and/or Black. Hate raid messages are found most commonly rooted in anti-Black racism and antisemitism. The authors document how these attacks elicited rapid community responses in both bolstering reactive moderation and developing proactive mitigations for future attacks. The study further discusses how platforms can better prepare for attacks and protect at-risk communities while considering the division of labor between community moderators, tool-builders, and platforms.

Within the context of content moderation in live streaming, HCI research has a critical role to play. The design and development of effective moderation systems that blend AI capabilities with human oversight can be an area of exploration. Besides, creating mechanisms to flag, report, and appeal decisions regarding negative content, ensuring user-friendly interfaces for all users, and studying how these mechanisms are used and can be improved, are all crucial areas of HCI research. HCI research can also contribute to designing more positive digital spaces by understanding user behaviors and incorporating measures that promote positive interactions. This could include encouraging users to reflect before posting or employing gamification techniques to reward positive behaviors (Seering et al., 2017). In the face of the negative content challenge, HCI researchers have a responsibility to help design systems that not only protect users from harm but also foster an atmosphere of respect and healthy communication. Efforts also need to be devoted to minimizing the psychological toll on human moderators exposed to harmful content. Investigating how job design, digital tools, and community support can mitigate this issue presents an interesting avenue for HCI research (Wohn, 2019).

One notable exploration in HCI to address content moderation in livestreaming is StoryChat (Yen et al., 2023), which investigates the influence of *proactive* moderation on viewers' engagement and prosocial behavior. StoryChat is a narrative-based viewer participation tool that utilizes a dynamic graphical plot to reflect chatroom negativity (Figure 13.3). When the chatroom's comments are negative, the visual narrative of StoryChat will also develop in a negative way. When viewers begin to post more positive comments, the visual narrative will turn more positive. StoryChat aims to enhance viewers' sense of community and engagement without constraining

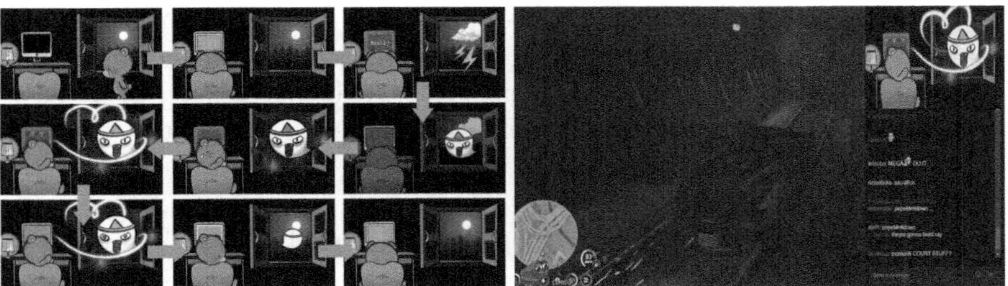

FIGURE 13.3 StoryChat, a live streaming viewer participation tool (Yen et al., 2023).

their participation and encourage them to actively participate in the ongoing discussion by sending prosocial comments in response to negative ones. The deployment study shows that StoryChat encourages viewers to contribute prosocial comments, increase viewer engagement, and foster viewers' sense of community. Viewers report a closer connection with streamers and other viewers because of the narrative design, suggesting that narrative-based viewer engagement tools have the potential to encourage community engagement and prosocial behaviors.

Content curation, on the other hand, deals with how content is presented and recommended to users. With an overflow of videos and live streams, providing users with relevant, useful, important, and personalized content is essential to maintaining user engagement. However, it inevitably raises concerns about fostering "echo chambers" and "filter bubbles." An essential part of HCI's role here is to explore ways to facilitate exposure to diverse content while still maintaining relevance and user interest.

Curation is an effective approach for community leaders to shape a space that matches their taste, norms, and values, but the practice is often intractable at social media scale due to the limited attention of curators. To address this issue, W. He et al. (2023) propose a transformer-based curation model called Cura that predicts whether each curator will upvote a post based on previous community upvotes. The model is applied to create a feed of content that it predicts the curator would want in the community. The curation model can accurately estimate the opinions of diverse curators and that changing curators for a community results in clearly recognizable shifts in the community's content. By leveraging Cura, the curation can reduce antisocial behaviors by half without extra moderation effort. This shows the potential of carefully designed curation models to help reduce negative behaviors online.

The significant role of HCI lies in ensuring an effective, affective, empathetic, and ethical approach to content moderation and curation at scale. The aim, ultimately, is to holistically enhance user experience, respect user rights, and foster a healthier and safer online environment for all.

The Rising Influence of Generative AI on Social Media

Generative AI, employing algorithms that can generate novel content, is catalyzing a transformative shift in the landscape of user-generated content, especially on short-video and live streaming platforms. From AI-generated music overlays on short videos to AI-powered avatars on live streams, the applications of generative AI are expanding. The introduction of AI into online content creation presents numerous challenges and opportunities for HCI. The first relates to the impact on user behaviors and engagement. Understanding how users perceive and interact with AI-generated content, compared to human-generated content, and how it affects their overall engagement is crucial.

For example, Jakesch et al. (2019) explore the potential impact of AI-mediated communication (AI-MC) on online self-presentation. The study takes a first look at how people perceive the trustworthiness of Airbnb hosts when they believe their profiles have been written by AI by conducting crowdsourced experiments. The study observes a new phenomenon that the authors' term the *Replicant Effect*, where participants mistrusted Airbnb hosts whose profiles were labeled as or suspected to be written by AI only when they thought they saw a mixed set of AI- and human-written profiles. The study results provide valuable insights into the potential impact of AI-MC on user perceptions of trustworthiness and highlight the importance of considering the role of AI in online self-presentation.

With the rise of large language models (LLMs) and their increasing applications in supporting writing, how LLMs may influence users is also a critical question. Jakesch et al. (2023) examine whether an opinionated language-model-powered writing assistant, which generates some opinions more often than others, impacts what users write with it and how they perceive this. The study conducted an online experiment where participants were asked to write a social media post discussing whether social media is good for society. The treatment group participants used a language-model-powered writing assistant that was configured to argue that social media is good or bad for society, making them write with opinionated agents. The study results showed that using the opinionated language model affected the opinions expressed in participants' writing and even shifted their opinions in the subsequent attitude survey. The study highlights that the opinions built into LLMs need to be monitored and engineered more carefully. The paper provides valuable insights into the potential impact of language-model-powered writing assistants on user opinions and highlights the importance of considering the role of AI in shaping user perceptions and attitudes.

A significant ethical issue relates to authenticating and crediting AI-generated content. *Deepfakes* represent a significant advancement in the field of generative AI. The term "deepfake" is a portmanteau of "deep learning" and "fake," reflecting the use of deep learning technology to create fake videos that are strikingly realistic (Gamage et al., 2022). Deepfakes leverage advanced machine learning and AI techniques to manipulate or fabricate visual and audio content with high precision, creating hyper-realistic videos. The increasing prevalence of deepfakes has raised serious concerns. The ability to fabricate realistic videos of people saying or doing things they did not poses significant risks to privacy, security, and public trust. It threatens to disrupt digital communication with potential misuse including the creation of non-consensual pornography, blackmail, political manipulation, and spreading disinformation.

Gamage et al. (2022) investigate how online communities on Reddit engage with deepfakes and the potential societal concerns that may arise from their use. The study analyzed deepfake conversations on Reddit from 2018 to 2021, including major topics and their temporal changes, as well as implications of these conversations. The study used a mixed-method approach, including topic modeling and qualitative coding, and found 6,638 posts and 86,425 comments discussing concerns of the believable nature of deepfakes and how platforms moderate them. The study also found that Reddit conversations were mostly pro-deepfake, and many users formed a community to create and share deepfake artifacts and built a marketplace regardless of the consequences. These findings further highlight the need for careful consideration of the role of AI in shaping online content and the potential impact of deepfakes on users.

In the context of HCI, developing mechanisms to indicate when content has been AI-assisted or entirely AI-generated, improving users' digital media literacy, and designing tools that can help users identify deepfakes can be necessary steps toward maintaining transparency and integrity on social media. Additionally, HCI research needs to explore design considerations for generative AI tools to ensure they are user-friendly and accessible. It should also focus on interfaces that allow users to seamlessly harness the power of AI in their content creation process without facing a steep learning curve. As generative AI continues to evolve, HCI researchers are presented with many emerging challenges and opportunities to shape the implementation of new generative AI algorithms to augment user capabilities and creativity, ensure ethical considerations, and ultimately, enhance the overall algorithmic experiences on social media.

From Challenges to Opportunities: Envisioning the Future of Social HCI

As the complexities of social media and online communities continue to evolve, especially in the spheres of short videos and live streaming, the challenges they present stand as valuable opportunities to push the boundaries of social HCI research.

First, there is an opportunity to holistically rethink and redesign user interfaces and user experience of short-video and live streaming platforms. As already suggested by several researchers in this area (Basapur et al., 2012; Hamilton et al., 2014; Lu, 2019), the move from textual to visual and interactive multi-modal communication will necessitate a shift in HCI models and theories. As these platforms continue to evolve and increasingly involving more diverse users, especially those who are traditionally overlooked in HCI research, engaging in iterative design processes and continually updating HCI models and theories to better understand and support the unique interaction paradigms they present will be crucial.

Second, the challenge of negative and inappropriate content within video-sharing platforms and the need for effective content moderation and curation at scale opens the door for HCI research make substantial contributions. Future HCI research can explore and design human-AI collaborative interfaces that not only empower users to report harmful or inappropriate content seamlessly (Fiesler et al., 2018), but also design and develop resilient systems that alleviate the burden on human moderators (Roberts, 2019; Wohn, 2019). Combining LLM-based AI algorithms with user feedback and advanced behavioral analysis, HCI researchers can create sophisticated content moderation tools that can more *proactively* moderate online communities (Yen et al., 2023). With the potential of AI-driven content moderation, HCI can help create a safer, more inclusive, and more positive online social environment for everyone.

Furthermore, considering the transformative potential of generative AI algorithms, future HCI research needs to focus not only on designing intuitive and accessible user interfaces but also on understanding and addressing the ethical implications of these algorithms (Dove et al., 2017; Gamage et al., 2022). This could include developing mechanisms for design transparency and ensuring that users are aware and in control when interacting with AI-generated content.

Moreover, as social media platforms become more and more immersive and interactive, future HCI research can investigate the physiological and psychological impacts of long-term usage of these platforms (Cummings & Bailenson, 2016). Designing user interfaces that take these factors into account will be key to the sustainable use of such social media platforms.

Finally, HCI has a crucial role in fostering inclusivity and accessibility in online communities. By designing accessible interfaces, studying and catering to diverse user needs, and creating safer and more respectful online spaces, future HCI research can help ensure that the benefits of these platforms are available to everyone (Bennett et al., 2018; Morris et al., 2016).

As social HCI looks toward the future, it must remain adaptable, embracing new trends and technologies, addressing emerging challenges, and continually renewing its commitment to creating user-centered, meaningful, and inclusive digital experiences.

REFERENCES

Basapur, S., Mandalia, H., Chaysinh, S., Lee, Y., Venkitaraman, N., & Metcalf, C. (2012). FANFEEDS: Evaluation of Socially Generated Information Feed on Second Screen as a TV Show Companion. In *Proceedings of the 10th European Conference on Interactive TV and Video*, 87–96. https://doi.org/10.1145/2325616.2325636

Bennett, C. L., E, J., Mott, M. E., Cutrell, E., & Morris, M. R. (2018). How Teens with Visual Impairments Take, Edit, and Share Photos on Social Media. In *Proceedings of the 2018 CHI Conference on Human Factors in Computing Systems*, 1–12. https://doi.org/10.1145/3173574.3173650

Cai, J., Wohn, D. Y., Mittal, A., & Sureshbabu, D. (2018). Utilitarian and Hedonic Motivations for Live Streaming Shopping. In *Proceedings of the 2018 ACM International Conference on Interactive Experiences for TV and Online Video*, 81–88. https://doi.org/10.1145/3210825.3210837

Chen, S., Chen, X., Lu, Z., & Huang, Y. (2023). "My Culture, My People, My Hometown": Chinese Ethnic Minorities Seeking Cultural Sustainability by Video Blogging. In *Proceedings of the ACM Human-Computer Interaction*, *7*(CSCW1). https://doi.org/10.1145/3579509

Chen, D. L., Freeman, D., & Balakrishnan, R. (2019). Integrating Multimedia Tools to Enrich Interactions in Live Streaming for Language Learning. In *Proceedings of the 2019 CHI Conference on Human Factors in Computing Systems*, 1–14. https://doi.org/10.1145/3290605.3300668

Cummings, J. J., & Bailenson, J. N. (2016). How Immersive Is Enough? A Meta-Analysis of the Effect of Immersive Technology on User Presence. *Media Psychology*, *19*(2), 272–309. https://doi.org/10.1080/15213269.2015.1015740

Dougherty, A. (2011). Live-Streaming Mobile Video : Production as Civic Engagement. In *Proceedings of the 13th International Conference on Human Computer Interaction with Mobile Devices and Services (MobileHCI '11)*, 425. https://doi.org/10.1145/2037373.2037437

Dove, G., Halskov, K., Forlizzi, J., & Zimmerman, J. (2017). UX Design Innovation: Challenges for Working with Machine Learning as a Design Material. In *Proceedings of the 2017 CHI Conference on Human Factors in Computing Systems*, 278–288. https://doi.org/10.1145/3025453.3025739

Fiesler, C., Jiang, J., McCann, J., Frye, K., & Brubaker, J. (2018). Reddit Rules! Characterizing an Ecosystem of Governance. In *Proceedings of the International AAAI Conference on Web and Social Media*, *12*(1). https://doi.org/10.1609/icwsm.v12i1.15033

Gamage, D., Ghasiya, P., Bonagiri, V., Whiting, M. E., & Sasahara, K. (2022). Are Deepfakes Concerning? Analyzing Conversations of Deepfakes on Reddit and Exploring Societal Implications. In *Proceedings of the 2022 CHI Conference on Human Factors in Computing Systems*. https://doi.org/10.1145/3491102.3517446

Gillespie, T. (2018). *Custodians of the Internet: Platforms, Content Moderation, and the Hidden Decisions That Shape Social media*. Yale University Press.

Glickman, S., McKenzie, N., Seering, J., Moeller, R., & Hammer, J. (2018). Design Challenges for Livestreamed Audience Participation Games. In *Proceedings of the 2018 Annual Symposium on Computer-Human Interaction in Play*, 187–199. https://doi.org/10.1145/3242671.3242708

Haimson, O. L., & Tang, J. C. (2017). What Makes Live Events Engaging on Facebook Live, Periscope, and Snapchat. In *Proceedings of the 2017 CHI Conference on Human Factors in Computing Systems (CHI'17)*, 48–60. https://doi.org/10.1145/3025453.3025642

Hamilton, W. A., Garretson, O., & Kerne, A. (2014). Streaming on twitch: Fostering Participatory Communities of Play Within Live Mixed Media. In *Proceedings of the 32nd Annual ACM Conference on Human Factors in Computing Systems (CHI'14)*, 1315–1324. https://doi.org/10.1145/2556288.2557048

Hamilton, W. A., Lupfer, N., Botello, N., Tesch, T., Stacy, A., Merrill, J. B., Williford, B., Bentley, F. R., & Kerne, A. (2018). Collaborative Live Media Curation: Shared Context for Participation in Online Learning. In *Proceedings of the 2018 CHI Conference on Human Factors in Computing Systems (CHI'18)*, Paper 555, 14 pages. https://doi.org/10.1145/3173574.3174129

Han, C., Seering, J., Kumar, D., Hancock, J. T., & Durumeric, Z. (2023). Hate Raids on Twitch: Echoes of the Past, New Modalities, and Implications for Platform Governance. In *Proceedings of the ACM Human-Computer Interaction*, 7(CSCW1). https://doi.org/10.1145/3579609

He, W., Gordon, M. L., Popowski, L., & Bernstein, M. S. (2023). Cura: Curation at Social Media Scale. In *Proceedings of the ACM on Human-Computer Interaction*, *7*(CSCW2). https://doi.org/10.1145/3610186

He, C., He, L., Lu, Z., & Li, B. (2023). Seeking Love and Companionship through Streaming: Unpacking Livestreamer-Moderated Senior Matchmaking in China. In *Conference on Human Factors in Computing Systems - Proceedings*, 18. https://doi.org/10.1145/3544548.3581195

Jakesch, M., Bhat, A., Buschek, D., Zalmanson, L., & Naaman, M. (2023). Co-Writing with Opinionated Language Models Affects Users' Views. In *Proceedings of the 2023 CHI Conference on Human Factors in Computing Systems*. https://doi.org/10.1145/3544548.3581196

Jakesch, M., French, M., Ma, X., Hancock, J. T., & Naaman, M. (2019). AI-Mediated Communication: How the Perception that Profile Text Was Written by AI Affects Trustworthiness. In *Proceedings of the 2019 CHI Conference on Human Factors in Computing Systems*, 1–13. https://doi.org/10.1145/3290605.3300469

Lessel, P., Vielhauer, A., & Krüger, A. (2017). Expanding Video Game Live-Streams with Enhanced Communication Channels: A Case Study. In *Proceedings of the 2017 CHI Conference on Human Factors in Computing Systems*, 1571–1576. https://doi.org/10.1145/3025453.3025708

Lin, J., & Lu, Z. (2017). The Rise and Proliferation of Live-Streaming in China: Insights and Lessons. In *International Conference on Human-Computer Interaction*, 632–637.

Lu, Z. (2019). Live Streaming in China for Sharing Knowledge and Promoting Intangible Cultural Heritage. *Interactions*, *27*(1), 58–63.

Lu, Z., Annett, M., Fan, M., & Wigdor, D. (2019). I Feel It Is My Responsibility to Stream: Streaming and Engaging with Intangible Cultural Heritage through Livestreaming. In *Proceedings of the 2019 CHI Conference on Human Factors in Computing Systems*, 229. https://doi.org/10.1145/3290605.3300459

Lu, Z., Habib, R., Wei, L.-Y., Dontcheva, M., Karahalios, K., Kazi, R. H., Wei, L.-Y., Dontcheva, M., & Karahalios, K. (2021). StreamSketch: Exploring Multi-Modal Interactions in Creative Live Streams. In *Proceedings of the ACM on Human-Computer Interaction*, *5*(CSCW), 1–26.

Lu, Z., Heo, S., & Wigdor, D. (2018a). StreamWiki: Enabling Viewers of Knowledge Sharing Live Streams to Collaboratively Generate Archival Documentation for Effective In-Stream and Post-Hoc Learning. In *Proceedings of the ACM on Human-Computer Interaction*, *2*(CSCW), Article 112, 24 pages. https://doi.org/10.1145/3274381

Lu, X., & Lu, Z. (2019). Fifteen Seconds of Fame: A Qualitative Study of Douyin, A Short Video Sharing Mobile Application in China. *Lecture Notes in Computer Science (Including Subseries Lecture Notes in Artificial Intelligence and Lecture Notes in Bioinformatics), 11578 LNCS*, 233–244. https://doi.org/10.1007/978-3-030-21902-4_17#Tab1

Lu, X., Lu, Z., & Liu, C. (2020). Exploring TikTok use and non-use practices and experiences in China. *Lecture Notes in Computer Science (Including Subseries Lecture Notes in Artificial Intelligence and Lecture Notes in Bioinformatics), 12195 LNCS*, 57–70. https://doi.org/10.1007/978-3-030-49576-3_5

Lu, Z., Xia, H., Heo, S., & Wigdor, D. (2018b). You Watch, You Give, and You Engage: A Study of Live Streaming Practices in China. In *Proceedings of the 2018 CHI Conference on Human Factors in Computing Systems (CHI'18)*, Paper 466, 13 pages. https://doi.org/10.1145/3173574.3174040

Morris, M. R., Zolyomi, A., Yao, C., Bahram, S., Bigham, J. P., & Kane, S. K. (2016). "With Most of It Being Pictures Now, I Rarely Use It": Understanding Twitter's Evolving Accessibility to Blind Users. In *Proceedings of the 2016 CHI Conference on Human Factors in Computing Systems*, 5506–5516. https://doi.org/10.1145/2858036.2858116

Preece, J. (2000). Online Communities: Designing Usability, Supporting Sociability. *Industrial Management & Data Systems*, *100*(9), 459–460. https://doi.org/10.1108/IMDS.2000.100.9.459.3

Reeves, S., Greiffenhagen, C., Flintham, M., Benford, S., Adams, M., Row Farr, J., & Tandavantij, N. (2015). I'd Hide You: Performing Live Broadcasting in Public. In *Proceedings of the 33rd Annual ACM Conference on Human Factors in Computing Systems*, 2573–2582. https://doi.org/10.1145/2702123.2702257

Roberts, S. T. (2019). *Content Moderation in the Shadows of Social Media*. Yale University Press. https://doi.org/10.2307/j.ctvhrcz0v

Seering, J., Kraut, R., & Dabbish, L. (2017). Shaping Pro and Anti-Social Behavior on Twitch through Moderation and Example-Setting. In *Proceedings of the 2017 ACM Conference on Computer Supported Cooperative Work and Social Computing*, 111–125. https://doi.org/10.1145/2998181.2998277

Tang, N., Tao, L., Wen, B., & Lu, Z. (2022). Dare to Dream, Dare to Livestream: How E-Commerce Livestreaming Empowers Chinese Rural Women. In *Proceedings of the 2022 CHI Conference on Human Factors in Computing Systems.* https://doi.org/10.1145/3491102.3517634

Tang, J. C., Venolia, G., & Inkpen, K. M. (2016). Meerkat and Periscope: I Stream, You Stream, Apps Stream for Live Streams. In *Proceedings of the 2016 CHI Conference on Human Factors in Computing Systems (CHI'16)*, 4770–4780. https://doi.org/10.1145/2858036.2858374

Wang, P., Hu, S., Wen, B., & Lu, Z. (2024). "There Is a Job Prepared for Me Here": Understanding How Short Video and Live-Streaming Platforms Empower Ageing Job Seekers in China. In *Proceedings of the CHI Conference on Human Factors in Computing Systems (CHI'24), May 11–16, 2024, Honolulu, HI, USA, 1.* https://doi.org/10.1145/3613904.3642959

Wohn, D. Y. (2019). Volunteer Moderators in Twitch Micro Communities: How They Get Involved, the Roles They Play, and the Emotional Labor They Experience. In *Proceedings of the 2019 CHI Conference on Human Factors in Computing Systems* (pp. 1–13). Association for Computing Machinery. https://doi.org/10.1145/3290605.s3300390

Wu, Q., Sang, Y., Wang, D., & Lu, Z. (2023). Malicious Selling Strategies in Livestream E-Commerce: A Case Study of Alibaba's Taobao and ByteDance's TikTok. *ACM Transactions on Computer-Human Interaction, 30*(3). https://doi.org/10.1145/3577199

Yang, S., Lee, C., Shin, H. V., & Kim, J. (2020). Snapstream: Snapshot-Based Interaction in Live Streaming for Visual Art. In *Proceedings of the 2020 CHI Conference on Human Factors in Computing Systems*, 1–12. https://doi.org/10.1145/3313831.3376390

Yen, R., Feng, L., Mehra, B., Pang, C. C., Hu, S., & Lu, Z. (2023). StoryChat: Designing a Narrative-Based Viewer Participation Tool for Live Streaming Chatrooms. In *Proceedings of the 2023 CHI Conference on Human Factors in Computing Systems.* https://doi.org/10.1145/3544548.3580912

14 Digital Wellbeing

Yasuyuki Hayama and Harshit Desai

INTRODUCTION

The Impact of Digital Technology on Wellbeing

In our increasingly digitized society, there is a burgeoning interest in the domain of digital-related wellbeing, often interchangeably referred to as digital wellness. This phenomenon encapsulates the intricate impact of digital technology and the Internet on an individual's holistic health and contentment, delving into the nuanced interplay of digital devices, social media, and online platforms. This comprehensive examination extends to their influence on physical and mental wellbeing, social connections, and the overall quality of life (Burr et al., 2020).

While technology bestows numerous advantages upon its users and society at large, a substantial body of literature highlights that users may experience negative emotions and significant breakdowns in self-regulation due to the excessive use of digital devices (e.g. Chakraborty et al., 2010; Hiniker et al., 2016; Cho et al., 2021; Abeele et al., 2022). Indeed, an overreliance on digital devices can be linked to adverse effects on mental health (Lanaj et al., 2014) and social interaction (Lee et al., 2014). For instance, mobile device use may disrupt the introspective processes inherent in in-person social interaction (Ko et al., 2016), impacting the quality of face-to-face conversations and relationships (Gergen, 2002). Furthermore, various studies demonstrate that devices such as smartphones can become sources of distraction, interfering with daily activities and ongoing tasks, such as studying, working, and driving (Harmon and Mazmanian, 2013). Distractions diminish users' productivity (Marino et al., 2018) and elevate stress levels (Mark et al., 2014), stemming from external stimuli like notifications and internal stimuli, such as the compulsion to check for new incoming emails. In India, where concerns about screen-based media and its impact on children's screen time have been raised, scholars propose the need to develop guidelines related to ensuring digital wellness and regulating screen time in infants, children, and adolescents (Gupta et al., 2022).

Building upon these insights, a prior study articulates digital wellbeing as a subjective, individualized experience representing an optimal balance between the advantages and disadvantages derived from mobile connectivity (Vanden Abeele, 2021). This experiential state involves affective and cognitive assessments of the integration of digital connectivity into daily life. Digital wellbeing is achieved when individuals navigate a state of maximal controlled pleasure and functional support, coupled with minimal loss of control and functional impairment. Similarly, an individual's "digital wellness" is characterized by the healthy engagement, both physically and mentally, with digital technology (McMahon and Aiken, 2015). McMahon and Aiken (2015) define a person's digital wellness as an indication of how healthily they relate to digital technology, both physically

DOI: 10.1201/9781032693606-18

and mentally. In contrast to prevailing definitions of e-health that predominantly accentuate positivity (Mackert et al., 2014), they underscore that adopting an "everything in moderation" approach is a more realistic stance toward technology usage. In essence, existing literature suggests maintaining an optimal balance between the benefits and drawbacks derived from digital technology.

Potential of Design for a Better Digital Wellbeing

Scholars acknowledge the potential problem areas in design to explore the capacity for digital wellbeing, categorizing them into medicalized, user-oriented, or design-oriented domains (Cecchinato et al., 2019). Historically, numerous consumer-focused technologies have aimed to maximize user engagement with their products and services. More recently, many technology companies have integrated digital wellbeing features, addressing aspects such as time management and encouraging breaks in usage. These developments occur against the backdrop of, and likely in response to, renewed concerns in the media regarding technology dependency and even addiction. The promotion of technology abstinence is also becoming increasingly widespread, evident, for example, through digital detox initiatives. Given that digital technologies are a crucial and valuable aspect of many people's lives, digital wellbeing features are arguably more preferable to abstinence (Cecchinato et al., 2019).

The potential for a range of design opportunities to establish an appropriate balance of advantages and disadvantages related to contact with digital devices is less contentious. However, we contend that a more nuanced and open perspective is necessary to consider the relationship between digital devices and people's wellbeing and to design a new world for the future. Indeed, recognizing the unintentional impact of technology on human wellbeing, Peters et al. (2018) emphasized a larger question: How can technology be designed to support wellbeing that encompasses more than just immediate hedonic experience but also its longer-term eudaimonia, or true flourishing? (Peters et al., 2018; Ryan and Deci, 2001, 2017; Sirgy, 2012). Therefore, this chapter delves into future design possibilities regarding "a better digital wellbeing," which prompts more meaningful interactions, specifically addressing how we can channelize the potential of this technology to foster improved habits and wellbeing in users' lives.

METHODOLOGY: CASE STUDY

This chapter scrutinizes various case studies and extracts implications for designing experiences aimed at fostering improved digital wellbeing. Specifically, adopting a hypothetical approach, we reflect on the existing knowledge concerning factors that demonstrably enhance health and wellbeing, while contemplating the prospect that digital technologies can contribute to wellbeing in a complementary or harmonious manner. Two potential enhancers, selected as representative examples from existing research, are deemed particularly effective in improving wellbeing: "connection to nature" (Wolsko and Lindberg, 2013) and "gamification" (Johnson et al., 2016)."

In the first perspective, "connection to nature," Wolsko and Lindberg (2013) expounded that exposure to nature correlates with numerous specific advantages in cognitive, emotional, and social functioning. Individuals actively participating in outdoor activities with appreciation are more likely to develop a strong connection with nature and experience elevated psychological wellbeing.

Another perspective under consideration is "gamification." Johnson et al. (2016) reviewed existing literature, asserting that the current evidence strongly supports the positive impact of gamification on health and wellbeing, particularly in influencing health behaviors.

Grounded in these two hypothetical perspectives, two case studies were undertaken as collaborative industry projects based on the action research approach. This approach involves a partnership between a client and an action researcher collaborating to diagnose a problem and construct a solution based on the diagnosis (Adeoye, 2023; Hayama and Desai, 2022).

Case 1: "Connection with Nature": Samsung India – Digital Wellbeing Project

The initial case serves as an illustrative exemplar of a practical design project undertaken through an industry-university collaboration at the MIT Institute of Design (MIT ID) in Pune, India. The objective of this project was to scrutinize the contemporary landscape of digital wellbeing among the youth in India. Simultaneously, the endeavor aimed to conceive innovative concepts that harness the potential of digital technology and systems, thereby fostering improved habits and wellbeing in the users' lives. A comprehensive delineation of the project's parameters is presented in Table 14.1.

The formulation of the principal challenge emanated from a thoughtfully crafted design brief put forth by the collaborative industry partner. Renowned as a

TABLE 14.1
Detailed Project Settings "Digital Wellbeing" Samsung India-MIT ID

Items	Settings
Goal	Proposing a novel system (framework) designed to enhance people's wellbeing in their daily lives through the use of digital devices in a harmonious manner.
Design Brief	To design a system (framework) using Samsung Galaxy Ecosystem which will bring wellbeing in end user's everyday life
Participants	3 master students in design management program 3 mentors from Samsung 2 faculties from MIT ID
Project details	• Duration: 6 months lasting industry-university collaborative project • Project approach: Project-Based Learning (PBL) • Location: Pune, India
Project activities	• Research: secondary research and primary research • Ideation • Concept development
Results	• Research results: Primary and secondary research results • Concept of new system that enhance people's digital wellbeing

leading Asian entity specializing in digital devices, the company exhibited a keen interest in discerning the nexus between digital devices and individual wellbeing. A pertinent illustration of a situation where technology has exhibited adverse effects is the pervasive overuse of social media in daily life. In this context, digital wellness transcends mere disconnection from devices; it encompasses a nuanced understanding of how digital devices are perceived and treated vis-à-vis people's wellbeing.

In alignment with these conceptual underpinnings, a design brief was articulated for the development of a system (framework) utilizing the Samsung Galaxy Ecosystem. The envisioned outcome was a system that contributes to the wellbeing of end-users in their day-to-day lives. The overarching goal of the project was to introduce a groundbreaking system (framework) explicitly tailored to augment the wellbeing of individuals through the judicious utilization of digital devices, ensuring a harmonious integration into their daily routines.

Implementation of the project involved the selection of four students from the design management program, chosen for their aptitude in crafting integrated product-service systems through a series of innovative design approaches. Facilitating and guiding the students in their project were two faculty members, the authors of this work, alongside three mentors affiliated with the collaborating company. The project transpired over a duration of six months, adopting the pedagogical approach of Project-Based Learning (PBL), and was executed through a hybrid online-offline mode in Pune, India.

Case 2: "Gamification" ForHealth – "Physical-digital Unified Gamification of Rehabilitation"

The second is an industry-academia collaboration between ForHealth, a venture company that creates innovative rehabilitation experiences by fusing the digital and the physical, and is a pioneering startup specializing in biomechanical technology with a focus on promoting musculoskeletal and neurological fitness. It is a pioneering startup company. The main goal of this startup is to revolutionize the field of rehabilitation and fitness by designing intelligent robotic assistive devices. At the same time, for the innovative company, it is expected that through a strong integration between academia and industry, students will be able to apply their theoretical learning to real-time practice, adding value to ForHealth's mission.

The goal of this project is to design and develop an interactive video game for physical therapy patients. This interactive game will make therapy engaging and motivate patients to continue their rehabilitation journey. A comprehensive definition of the project parameters is provided in Table 14.2.

The starting point for this project was the hypothesis that incorporating digital and physical gamification into treatment could increase patient engagement in physical therapy and, in turn, contribute to a wellbeing treatment experience for patients. ForHealth's device, which stands apart from others, has the unique ability to learn and reproduce movement patterns, ensuring the fulfillment of recommended repetitive exercises, a shortcoming of traditional physical therapy methods. This cost-effective solution provides passive range-of-motion exercises and adaptable

TABLE 14.2
Detailed Project Settings

Items	Settings
Goal	To explore the role of gamification and innovative digital engagement tools in enhancing the rehabilitation and recovery of physiotherapy patients.
Design Brief	To design a digital engagement tool embedded in an Intelligent LowerLimb Robotic device to enhance the physiotherapy treatment process.
Participants	3 master students in design management program 2 students from game design and development program 2 faculties from MIT ID 2 mentors from ForHealth
Project details	• Duration: 6 months lasting industry-university collaborative project • Project approach: Industry-Integrated Learning program • Location: Pune, India
Project activities	• Research: secondary research and primary research • Ideation, Design • Game development, software engineering and testing
Results	• Research results: Primary and secondary research results • Deployment of interactive games in the Intelligent LowerLimb Robotic device for physiotherapy treatment

resistance for active exercises. The device's portability also allows it to provide telemedicine, allowing patients to receive home therapy under supervised or unsupervised conditions. By incorporating gamification into treatment and collecting real-time data, ForHealth believed it could increase patient engagement and monitor progress. In addition, the technology optimizes therapist efficiency by allowing them to treat multiple patients simultaneously, potentially remaking the physical therapy industry into a more scalable model.

With these hypotheses in mind, three basic questions were set for the project.

1. Do patients need to feel motivated to complete physical therapy?
2. Should sessions be more interactive and engaging?
3. Can games accomplish the above?

The expected outcome is the conceptualization of a gamified digital experience utilizing ForHealth's intelligent lower limb robotic device.

Five students from the Design Management and Game Design programs were selected to implement the project, each with an aptitude for creating innovative digital experiences. The students' projects were facilitated and guided by two faculty members, including one of the authors of this paper, and two mentors from collaborating companies. The project adopted an Industry-Integrated Learning Program in Pune, India, over a period of six months.

After independently analyzing each case, the discussion section will extract and synthesize common insights from the two cases. The analytical perspective is centered on digital wellbeing, aiming to provide a comprehensive understanding.

The objective is to adopt a broader view of digital wellbeing and draw implications for human wellbeing through digital technology from a future-oriented standpoint.

RESULTS

Case 1: Samsung India, Digital Wellbeing Project

The first case of Samsung India exemplifies an ambitious initiative, illustrating how design can proffer alternative and visionary perspectives on digital wellbeing, thereby fostering increased human happiness through an industry-university collaborative workshop. The outcomes elucidate the comprehensive research findings of the project, encapsulating both the summative results and the evolved conceptual framework derived from the design endeavor.

Research Results

During the research phase, the design team systematically explored design opportunities and delineated their design trajectory through a combination of secondary research and primary research. The secondary research phase involved a comprehensive examination of the nuances of digital wellbeing. Concurrently, in the primary research segment, the team conducted a diary study (Adler et al., 1998) to delve into both the peak moments and nadirs in the daily lives of the target demographic. Ultimately, the design direction crystallized from these investigations.

Guided by the unequivocal premise that "wellbeing encompasses the amalgamation of feeling good and functioning effectively; the encounter with positive emotions such as happiness and contentment, coupled with the unfolding of one's potential, exerting control over one's life, harboring a sense of purpose, and fostering positive relationships" (Huppert, 2009; Ruggeri et al., 2020), the primary research sought to comprehend the stimuli responsible for evoking specific emotional responses within the target groups. The findings elucidate three primary catalysts for moments of happiness: "1. Care (someone thinking about you)," "2. Quality time (with self and others)," and "3. Initiative (making a decision)." Conversely, the triggers for low points were identified as the top four factors: "1. Overwhelmed (having too much on the mind)," "2. Out of routine (unanticipated events)," "3. Lonely (feeling alone)," and "4. Unproductive & Tired."

Building upon the comprehensive insights gleaned from the entire research endeavor, the team distilled their design direction into a dual-faceted approach to wellbeing. Specifically, on one front, the inquiry focused on elucidating how digital devices and technology could elevate the peak moments in the lives of the target demographic. Conversely, on the other front, it addressed the question of how these technologies could mitigate negative emotions. In a tangible sense, regarding the promotion of positive emotions, the design direction aspired to nurture individuals by incorporating considerate elements that evoke feelings of happiness and affection. Additionally, it sought to ensure connectivity to people and entities that facilitate the quality utilization of time, as well as motivate individuals to make decisions that yield satisfaction and confidence.

In terms of mitigating negative emotions, the design direction rigorously aimed to prevent the exacerbation of detrimental feelings, such as overwhelm, deviation from routine, isolation and loneliness, unproductivity, and fatigue.

Conception: "Nature-Embracing Digital Wellbeing Experience: Soulo"

Reflecting the research results and basic understanding of digital wellbeing, the design team developed a new concept which intends to change the meaning of it radically. After iterative ideation and intensive discussion among the design team and company's participants and faculties, the design team defined the designing situation on "outdoor activity and wellbeing."

Extant literature suggests that the mere presence of a natural environment can have significant positive effects on wellbeing (e.g., Berman et al., 2008; Van den Berg et al., 2010). Furthermore, Wolsko and Lindberg (2013) examined that those types of experiences that explicitly generate genuine appreciation of and connection with nature are likely to be the most beneficial for psychological wellbeing. They also stressed that mental health should be enhanced by interventions that facilitate appreciative outdoor recreational activities, as well as those that promote mindfulness (Wolsko and Lindberg, 2013).

Based on the assumptions on the relationship between natural experiences and wellbeing, the design team framed the How Might We (HMW) question; "How might we encourage people to spend more time outdoors so that it has a positive impact on their overall wellbeing?" To develop solutions and realize the specific solution, assumption was set up; to empower the target persona's mindfulness and wellbeing by harnessing the Samsung ecosystem, seamlessly integrating phone, smartwatch, and earphones for a holistic wellness experience. This design inquiry is grounded in the fundamental premise that digital devices play a pivotal role in enhancing individuals' wellbeing by providing an avenue for them to connect with and immerse themselves in nature. Beyond this foundation, the inquiry postulates the prospect that digital devices possess the potential to broaden and amplify the experiential dimensions of nature. The overarching aim is to facilitate the extension of wellbeing benefits derived from individuals' active engagement with nature through the augmentation afforded by digital devices.

The ultimate objective of the final concept is to offer a wholly novel digital wellbeing experience rooted in nature, aptly named "Soulo." Conceived as a spiritual entity, Soulo embodies a guiding force designed to reconnect individuals with the rejuvenating embrace of nature. The mission of Soulo is to subtly encourage individuals to venture outdoors, characterized by an infectious zest for life, replete with enthusiasm and curiosity. Functioning as a benevolent guide, Soulo gently prompts users to take respite from their digital routines and immerse themselves in the outdoors. It fosters a connection to nature by thriving on sunlight, fresh air, and the Earth's energy. Soulo grows symbiotically with users' increased engagement with nature, its aura expanding with each step taken outside. This growth symbolizes users' commitment to embracing outdoor experiences, marking a shared journey toward reconnecting with the natural world.

The overarching aim is to develop an outdoor application that motivates individuals to spend more time in natural settings, employing a digital soul that

FIGURE 14.1 Soulo application image, the character represents the user's digital soul.

encapsulates their outdoor spirit conditions (refer to Figure 14.1). The primary function entails providing a timer to display the duration users have spent outdoors and gently encouraging them through notifications (refer to Figure 14.2). Through this functionality, the Soulo application actively engages users in spending quality time with both loved ones and nature. As depicted in Figures 14.3 and 14.4, the basic screens for both the smartphone and smartwatch interfaces are intentionally crafted to feature empathetic soul characters, each expressing facial expressions reflective of users' emotional and spiritual states.

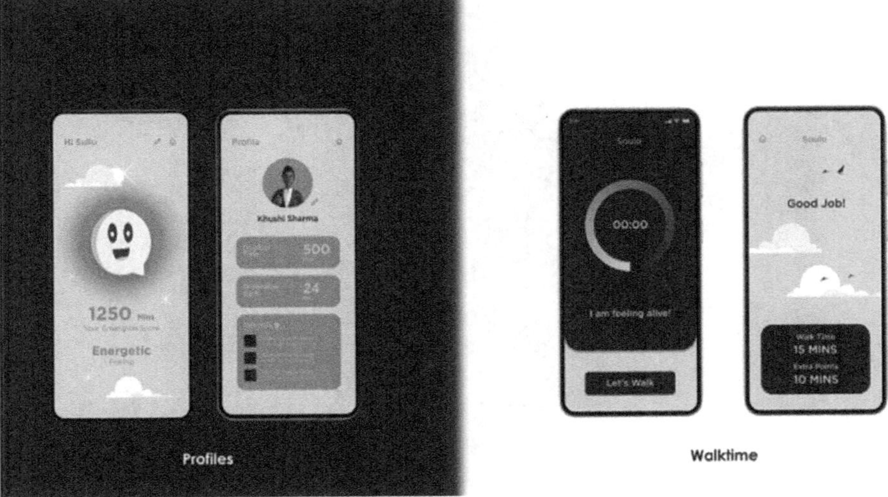

FIGURE 14.2 The basic function to provide a timer and to nudge the user through notifications.

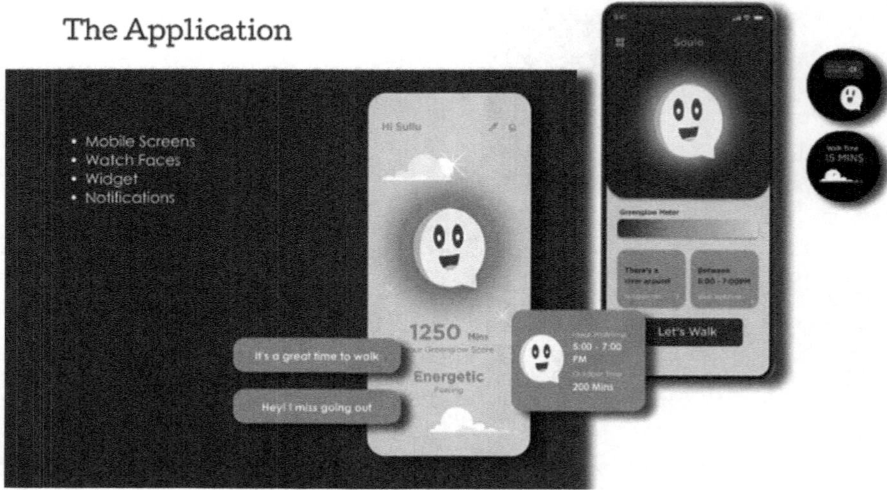

FIGURE 14.3 The basic screens of the application in smartphones.

Beyond its encouragement of outdoor activities, the Soulo application offers additional user experiences through discovery functions. It motivates users to forge new connections by exploring outdoor experiences with friends and colleagues, while also facilitating the creation of a memory journal (refer to Figure 14.5) to capture and preserve their unique outdoor journeys. Ultimately, Soulo's primary value proposition lies in facilitating digital breaks and aiding users in discovering the wonders of nature and their immediate surroundings.

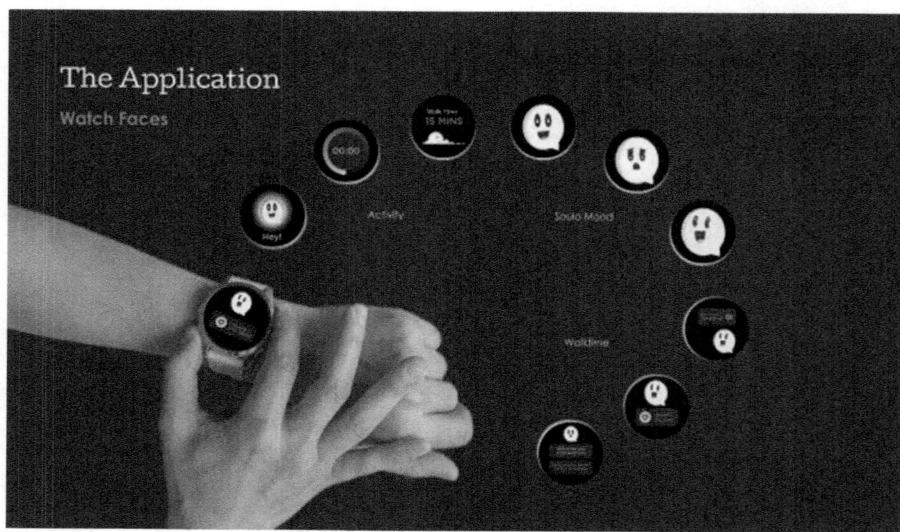

FIGURE 14.4 Smartwatch faces.

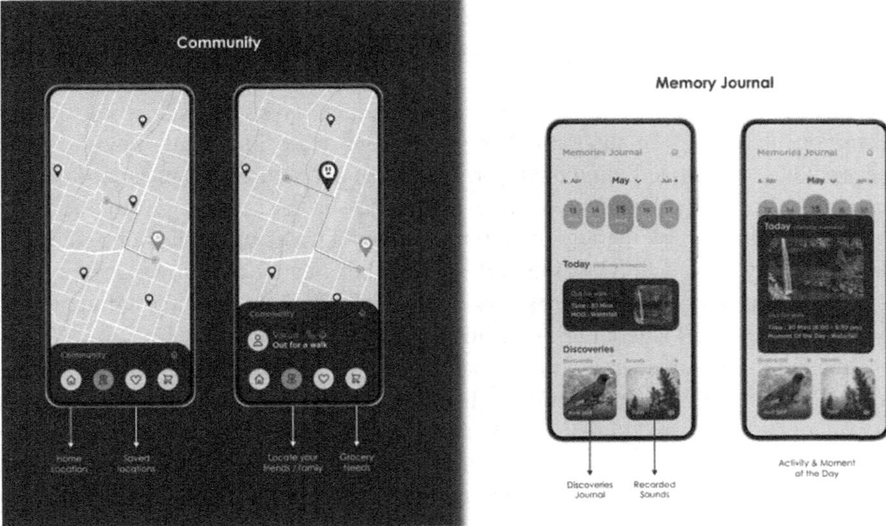

FIGURE 14.5 Discovery functions: community and memory journal.

The Soulo concept endeavors to redefine the conventional notion of digital wellbeing, particularly in relation to the equilibrium between time spent engaging with digital devices. Put differently, it actively encourages users to proactively immerse themselves in the natural world through the intermediary of digital devices and digital experiences.

Moreover, the Soulo concept serves the purpose of augmenting the influence and potency of nature on wellbeing. The objective extends beyond merely maximizing exposure to nature; it seeks to prolong and enrich the encounter with nature through the capabilities of digital devices.

In this context, the digital wellbeing proposition presented by Soulo introduces an entirely novel dimension that not only aligns with but also enhances the overall experience of nature.

Case 2: ForHealth – A Unified Physical-Digital Gamification Approach to Rehabilitation

The case of ForHealth offers insights into the potential of a unified digital-physical gamification approach for enhancing wellbeing. This collaborative industry-academia project, involving a venture firm and a university, initiated by exploring the prospects of combining digital and physical rehabilitation through gamification to develop a novel experiential concept.

Research Results

Throughout the research phase, the design team conducted a comprehensive blend of secondary and primary research to explore the potential of providing gamified experiences to enhance wellbeing in physical therapy. Through interviews encompassing diverse demographics – physical therapists, patients, gamers, and game developers – the findings hypothesized in the secondary research were further elucidated.

To distill the findings, despite variations in patient types and symptom severity, common challenges in physical therapy sessions were identified. Physiotherapy necessitates specialized equipment for leg movement, and exercises are conducted under medical supervision or at home with patient assistance. Given the often protracted duration of therapy, maintaining ongoing patient motivation is crucial. Consequently, gamification was hypothesized to be effective in four ways: 1. presenting challenges and engagement, 2. fostering creativity, 3. motivating individuals to achieve more, and 4. encouraging additional effort.

Indeed, existing research, as concluded by Johnson et al. (2016), indicates widespread acceptance of gamification in health and wellbeing. A literature review of pertinent studies revealed that gamification consistently yields positive impacts on health and wellbeing. Positioned as a promising alternative in comparison to existing approaches like serious games and persuasive health technologies, gamification embodies a "new model for health," characterized as a "fascinating, ubiquitous, lifelong health interface across the lifespan" (Sawyer, 2014). Proponents highlight seven potential benefits: (1) supporting intrinsic motivation, (2) serving as a "seductive, ubiquitous, and lifelong health interface" for mobile technology and ubiquitous sensors, (3) appealing to a broad audience, (4) being applicable to a wide range of health risks and factors, (5) cost-effectiveness in extending existing systems, (6) integrating seamlessly into everyday life, and (7) directly promoting wellbeing by providing positive experiences (Johnson et al., 2016).

Reflecting on the project's findings, the design team, drawing from interviews with physiotherapists and existing research on gamification, honed in on specific foot activities that gamification should target to aid patients in their recovery (Figure 14.6). Subsequent design activities were then directed toward conceptualizing ideas for developing hybrid solutions based on gamification, physical support with rehabilitation equipment, and game design utilizing digital technology.

Conception: "Physical-digital Unified Gamification of Rehabilitation"

Focusing on lower limb activities, the design teams innovatively devised new experiential activities that seamlessly integrate physical rehabilitation exercises for the legs with digital gamification elements.

The foundational physical activities are intended to be facilitated by ForHealth's inaugural product, the "Rehub Buddy" (Figure 14.7). The Rehub Buddy is an Intelligent Lower Limb Orthosis designed to assist patients in various passive range-of-motion exercises (Figure 14.8). When required, it naturally resists patients during active isometric and isotonic exercises. Following collaboration and guidance from the physiotherapist during the initial session, this orthosis takes charge of repetitive exercises, faithfully replicating assisting and resisting patterns throughout the range of motion (Figure 14.9).

Concurrently, alongside the physical activities, digital experiences enhance treatment through gamification and also gather real-time data on the patient's vital

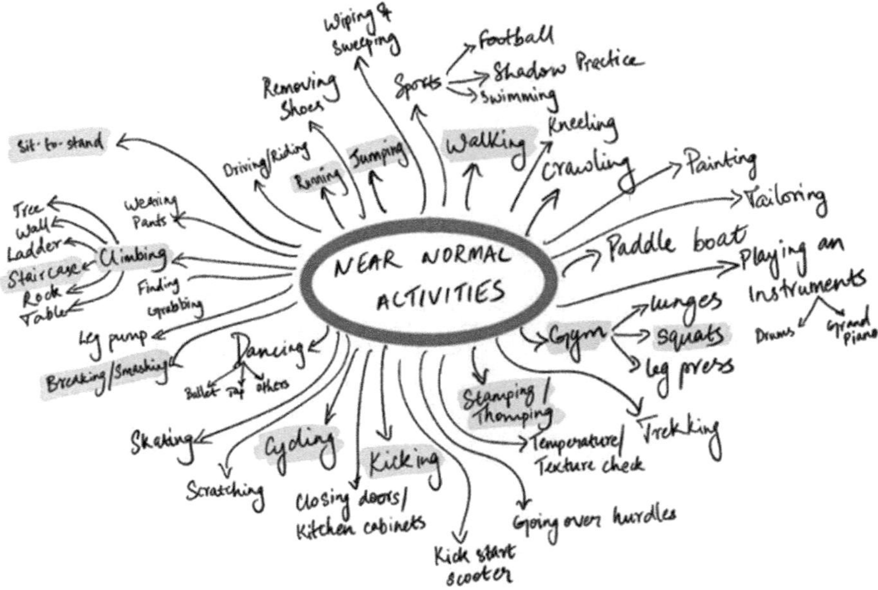

FIGURE 14.6 Brainstorming the near normal leg activities that gamification should focus on.

parameters. In this industry-university collaborative project, the fundamental components and concepts of gamification were explored, encompassing characters, storylines, and storyboards (Figure 14.10). After several intensive ideation workshops, the selected storyline revolves around a heartwarming concept. In this narrative, the player, representing the patient, interacts with the game by kicking logs to rescue a dog trapped underneath. The beauty of this narrative lies in its synchronization with the real-time movements of the patient on the physiotherapy device, creating an immersive and feel-good gaming experience.

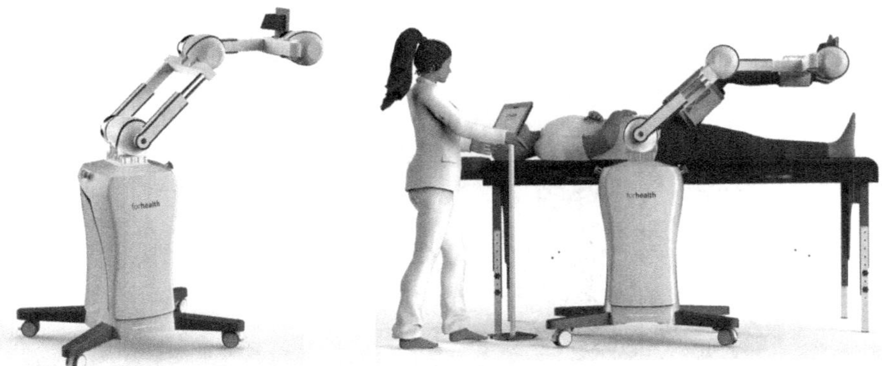

FIGURE 14.7 Rehub Buddy, the Intelligent Lower Limb Orthosis.

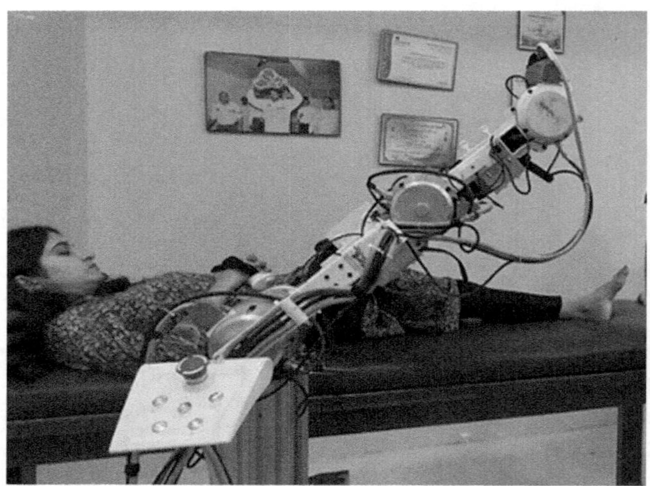

FIGURE 14.8 Healthy user testing with Rehab Buddy.

Navigating decisions on elements such as the environment, storyline details, and the choice between 2D and 3D graphics, collaborative efforts ensured a comprehensive understanding of the game's development trajectory. Decisions, such as opting for a 3D format to enhance realism, substituting logs with a dog house for ease of development, and choosing Unity software for its widespread use and seamless integration, highlighted the thoughtful considerations driving the development process (Figure 14.11).

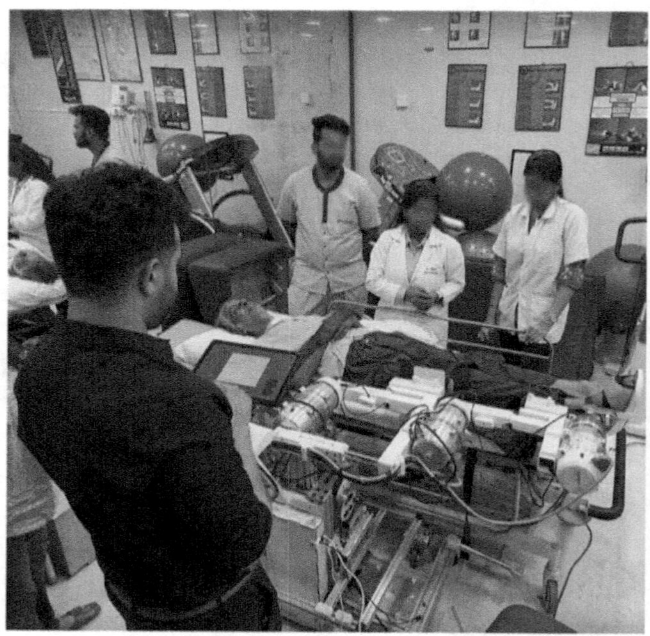

FIGURE 14.9 User playing game on the device.

FIGURE 14.10 Storyboard which represent the storyline of the game.

The outcome of this collaborative journey was a fully functional game developed to industry standards, seamlessly integrated into the company's Android application by the in-house developers. Rigorous testing with healthy users affirmed its interactivity and ability to keep users engaged throughout their physiotherapy sessions. Serving as a proof of concept, the game successfully mirrored real-time movements,

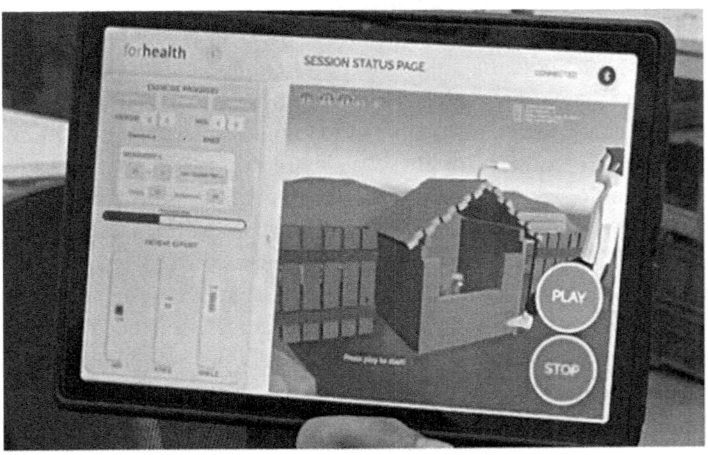

FIGURE 14.11 3D developed game.

validating its efficacy in tandem with their physiotherapy device for lower limb exercises. These interactive games make therapy engaging, motivating patients to stay committed to their rehabilitation journey.

Looking ahead, ForHealth is committed to expanding its game library to cater to diverse patient needs, incorporating exercises for both the upper limbs (such as shoulder and elbow exercises) and lower limbs (introducing more variations and stories in the game) and other functional exercises. This fruitful collaboration with MIT ID not only exemplifies the fusion of academic knowledge and industry expertise but also underscores the success of our shared mission in revolutionizing physiotherapy through innovative and engaging solutions.

The game stands as a testament to the potential of combining physical and digital wellbeing activities through a gamification approach. Hence, more meaningful interaction can be realized in healthcare technology.

SUMMARY OF THE TWO CASES

The two cases propose that digital technologies can contribute to enhancing people's wellbeing in distinct ways. The Samsung case implies that digital technologies may enhance wellbeing by playing a complementary or reinforcing role in fostering a "connection with nature." In contrast, the ForHealth case presents the potential of digital technology to improve wellbeing through gamification methods, coupled with physical exercise, thereby amplifying the efficacy of rehabilitation.

DISCUSSION

The aim of this chapter is to transcend prevailing perspectives on digital wellbeing and offer insights into the utilization of digital technology for enhanced wellbeing. Through two industry-academia collaboration case studies, novel facets of digital wellbeing are unveiled. The initial case study illustrates the potential success of employing digital technology to augment its efficacy in enhancing human wellbeing through nature-based experiences. Conversely, the second case study illuminates the prospect of leveraging digital technology to ameliorate the wellbeing of individuals within the rehabilitation context by means of direct intervention. This second case study underscores the potential synergies between physical and digital technologies in crafting experiences that amplify the effectiveness of health promotion treatments, notably rehabilitation, through methodologies such as gamification.

An overarching synthesis of these case studies posits that adeptly integrating digital technologies into conditions conducive to improved wellbeing, such as interactions with nature and gamification, can amplify the extent of these enhancements. This proposition implies an expansion beyond the conventional understanding of digital wellbeing, suggesting that it transcends the mere balancing act between the advantages and disadvantages in fostering a healthy relationship with digital devices. Rather, digital wellbeing is reconceptualized as the judicious utilization of digital technologies in conjunction with diverse environmental conditions and methodologies to establish favorable circumstances for human wellbeing.

Three principal contributions of this paper merit emphasis. First, it introduces a broader and more affirmative connotation of digital wellbeing, challenging the prevailing notion of seeking balance between advantages and disadvantages for human wellbeing. Second, it demonstrates that methodologies like practical action research by design (e.g., research-through-design) offer an alternative perspective on conventional conceptual definitions. Third, grounded in the newly hypothesized concept of "better digital wellbeing," it problematizes the problem domain of future design, envisaging the creation of products, services, and worldviews. This view can possibly open up a new "art" of digital experience design that encompasses the more emotional, and atmospheric elements of users' digital interactions such as attraction, seduction, and engagement (Leung, 2008).

From a design and futurological standpoint, this definition is rife with implications, presenting several promising avenues for further research. First, exploring the conditions under which digital technologies are conducive to enhancing people's wellbeing, defining the parameters of "better digital wellbeing." Engagement with nature and gamification are just initial examples, with numerous other conditions warranting consideration. Second, investigating the opportunities for designing new products and services under the conditions of better digital wellbeing, encapsulating the ethos of design for improved digital wellbeing. Last, envisioning the future trajectory of better digital wellbeing, offering a visionary perspective.

Finally, it is crucial to acknowledge the limitations of this chapter. First, concerning the research methodology, this study is based on a limited number of cases from industry-academia projects in India. Consequently, the validity of the insights derived necessitates further scrutiny and validation through diverse methodologies and across different countries and cultural contexts. Second, there is a limitation in the number of initial hypotheses, as this study primarily adopts contact with nature and gamification as existing conditions believed to enhance wellbeing. Recognizing the multitude of other factors that likely contribute to enhanced wellbeing, further research is imperative to explore a broader spectrum of conditions for better digital wellbeing.

ACKNOWLEDGMENT

We sincerely thank our industry-university collaborative partners and our students. Special thanks go to Anirudha Hajare, Sanika Sud, Abhijit Lade of Samsung India for collaboration and mentorship for our students and the project. We wish to express our gratitude to the entire team of ForHealth for intensive collaboration with our school.

REFERENCES

Abeele, M. M. V. (2021). Digital wellbeing as a dynamic construct. *Communication Theory, 31*(4), 932–955.

Abeele, M. M. V., Halfmann, A., & Lee, E. W. (2022). Drug, demon, or donut? Theorizing the relationship between social media use, digital well-being and digital disconnection. *Current Opinion in Psychology, 45*, 101295.

Adeoye, M. A. (2023). From variables to research design: A deep dive into educational research methodology. *Journal of Education Research and Evaluation*, *7*(4), 622–628.

Adler, A., Gujar, A., Harrison, B. L., O'hara, K., & Sellen, A. (1998). A Diary Study of Work-Related Reading: Design Implications for Digital Reading Devices. In *Proceedings of the SIGCHI Conference on Human Factors in Computing Systems* (pp. 241–248).

Berman, M. G., Jonides, J., & Kaplan, S. (2008). The cognitive benefits of interacting with nature. *Psychological science*, 19(12), 1207–1212.

Burr, C., Taddeo, M., & Floridi, L. (2020). The ethics of digital well-being: A thematic review. *Science and Engineering Ethics, 26*, 2313–2343. https://doi.org/10.1007/s11948-020-00175-8.

Cecchinato, M. E., Rooksby, J., Hiniker, A., Munson, S., Lukoff, K., Ciolfi, L., Thieme, A., & Harrison, D. (2019). Designing for Digital Wellbeing: A Research & Practice Agenda. In *Paper presented at the Extended Abstracts of the 2019 CHI Conference on Human Factors in Computing Systems* (pp. 1–8). ACM.

Chakraborty, K., Basu, D., & Kumar, K. V. (2010). Internet addiction: Consensus, controversies, and the way ahead. *East Asian Archives of Psychiatry, 20*(3), 123–132.

Cho, H., Choi, D., Kim, D., Kang, W. J., Choe, E. K., & Lee, S. J. (2021). Reflect, not regret: Understanding regretful smartphone use with app feature-level analysis. *Proceedings of the ACM on Human-Computer Interaction, 5*(CSCW2), 1–36.

Gergen, K. J. (2002). The Challenge of Absent Presence. *Perpetual Contact: Mobile Communication, Private Talk, Public Performance* (pp. 227–241). http://doi.org/10.1017/CBO9780511489471.018; https://works.swarthmore.edu/fac-psychology/569

Gupta, P., Shah, D., Bedi, N., Galagali, P., Dalwai, S., & Agrawal, S., … & IAP Guideline Committee on Digital Wellness and Screen Time in Infants, Children and Adolescents (2022). Indian Academy of pediatrics guidelines on screen time and digital wellness in infants, children and adolescents. *Indian Pediatrics, 59*(3), 235–244.

Harmon, E., & Mazmanian, M. (2013). Stories of the Smartphone in Everyday Discourse: Conflict, Tension & Instability. In *Proceedings of the SIGCHI Conference on Human Factors in Computing Systems* (pp. 1051–1060).

Hayama, Y., & Desai, H. (2022). Educational probe for developing online education: A case of online problem-based learning in design education in India. *Proceedings of the Design Society, 2*, 2283–2292.

Hiniker, A., Hong, S., Kohno, T., & Kientz, J. A. (2016). MyTime: Designing and Evaluating an Intervention for Smartphone Non-Use. In *Paper presented at the Proceedings of the 2016 CHI Conference on Human Factors in Computing Systems, San Jose, CA*.

Huppert, F. A. (2009). Psychological well-being: Evidence regarding its causes and consequences. *Applied Psychology: Health and Well-Being, 1*(2), 137–164.

Johnson, D., Deterding, S., Kuhn, K. A., Staneva, A., Stoyanov, S., & Hides, L. (2016). Gamification for health and wellbeing: A systematic review of the literature. *Internet Interventions, 6*, 89–106.

Ko, M., Choi, S., Yatani, K., & Lee, U. (2016, May). Lock n'LoL: Group-Based Limiting Assistance App to Mitigate Smartphone Distractions in Group Activities. In *Proceedings of the 2016 CHI Conference on Human Factors in Computing Systems* (pp. 998–1010).

Lanaj, K., Johnson, R. E., & Barnes, C. M. (2014). Beginning the workday yet already depleted? Consequences of late-night smartphone use and sleep. *Organizational Behavior and Human Decision Processes, 124*(1), 11–23.

Lee, U., Lee, J., Ko, M., Lee, C., Kim, Y., Yang, S., … & Song, J. (2014, April). Hooked on Smartphones: An Exploratory Study on Smartphone Overuse Among College Students. In *Proceedings of the SIGCHI Conference on Human Factors in Computing Systems* (pp. 2327–2336).

Leung, L. (2008). *Digital experience design: Ideas, industries, interaction* (pp. 1–128). Intellect.

Mackert, M., Champlin, S. E., Holton, A., Muñoz, I. I., & Damásio, M. J. (2014). eHealth and health literacy: A research methodology review. *Journal of Computer-Mediated Communication*, *19*(3), 516–528.

Marino, C., Gini, G., Vieno, A., & Spada, M. M. (2018). A comprehensive meta-analysis on problematic Facebook use. *Computers in Human Behavior*, *83*, 262–277. https://doi.org/10.1016/j.chb.2018.02.009

Mark, G., Wang, Y., & Niiya, M. (2014). Stress and Multitasking in Everyday College Life: An Empirical Study of Online Activity. In *Proceedings of the SIGCHI Conference on Human Factors in Computing Systems (Toronto, Ontario, Canada) (CHI'14)*. ACM, New York, NY, USA, 41–50. https://doi.org/10.1145/2556288.2557361

McMahon, C., & Aiken, M. (2015, October). Introducing Digital Wellness: Bringing Cyberpsychological Balance to Healthcare and Information Technology. In *2015 IEEE International Conference on Computer and Information Technology; Ubiquitous Computing and Communications; Dependable, Autonomic and Secure Computing; Pervasive Intelligence and Computing* (pp. 1417–1422). IEEE.

Peters, D., Calvo, R. A., & Ryan, R. M. (2018). Designing for motivation, engagement and wellbeing in digital experience. *Frontiers in Psychology*, *9*, 797. http://doi.org/10.3389/fpsyg.2018.00797

Ruggeri, K., Garcia-Garzon, E., Maguire, Á, Matz, S., & Huppert, F. A. (2020). Well-being is more than happiness and life satisfaction: A multidimensional analysis of 21 countries. *Health and Quality of Life Outcomes*, *18*(1), 1–16.

Ryan, R. M., & Deci, E. L. (2001). On happiness and human potentials: A review of research on hedonic and eudaimonic well-being. *Annual Review of Psychology*, *52*, 141–166. http://doi.org/10.1146/annurev.psych.52.1.141

Ryan, R. M., & Deci, E. L. (2017). *Self-Determination Theory: Basic Psychological Needs in Motivation, Development, and Wellness*. New York, NY: Guilford Press.

Sawyer, B., (2014). Games for Health. MIT. In *Presented at the MIT Media Lab Talks, Cambridge, MA*. https://www.media.mit.edu/video/view/wellbeing-2014-11-05.

Sirgy, J. (2012). *The Psychology of Quality of Life: Hedonic Well-Being, Life Satisfaction, and Eudaimonia*. New York, NY: Springer.

Van den Berg, A. E., Maas, J., Verheij, R. A., & Groenewegen, P. P. (2010). Green space as a buffer between stressful life events and health. *Social science & medicine*, 70(8), 1203–1210.

Wolsko, C., & Lindberg, K. (2013). Experiencing connection with nature: The matrix of psychological well-being, mindfulness, and outdoor recreation. *Ecopsychology*, *5*(2), 80–91. http://doi.org/10.1089/eco.2013.0008

Index

Note: **Bold** page numbers refer to **tables** and *Italic* page numbers refer to *figures*.

D

R

S

T

For Product Safety Concerns and Information please contact our EU representative GPSR@taylorandfrancis.com Taylor & Francis Verlag GmbH, Kaufingerstraße 24, 80331 München, Germany

Batch number: 10397790

Printed by Printforce, the Netherlands